Basic Hydraulics

Basic Hydraulics

Fluid Power Workhorse

Jay F. Hooper

Carolina Academic Press
Durham, North Carolina

Copyright © 2012 Jay F. Hooper
All Rights Reserved

Library of Congress Cataloging-in-Publication Data

Hooper, Jay F.
  Basic hydraulics : fluid power workhorse / Jay F. Hooper.
       p. cm.
  ISBN 978-1-59460-835-3 (alk. paper)
 1. Fluid power technology. 2. Hydraulic fluids.
I. Title.

  TJ843.H68 2010
  621.2--dc22   2010023557

Carolina Academic Press
700 Kent Street
Durham, NC 27701
Telephone (919) 489-7486
Fax (919) 493-5668
www.cap-press.com

Printed in the United States of America

Contents

Preface   xi

Introduction   xiii

Section One

1 · Hydraulic Fundamentals   3
    The Four States of Matter   3
    Change of State   3
    STP   5
    Normal Air   5
    Free Air   5
    Relative Humidity   6
    Dew Point   7
    FPS [MPS or mps or m/sec.]   8
    Scales   8
    Absolute Pressure Scale   8
    Gauge Pressure Scale   8
    Vacuum Pressure Scale   9
    Major Constituents of Air   9
    Compressibility   10

2 · Hydraulic Fluid Treatment or Conditioning   13
    Primary Fluid Treatment   13
    Strainers   13
    Suction Line Filters   14
    Pressure Line Filters   14
    Return Line Filters   14
    Hydraulic Fluid Cooler   14
    Hydraulic Fluid Heater   14
    Separation of Water   15
    Separation of "Big" Dirt   15
    Breather Adsorption Filter   15

Contents

	Micron	16
	Secondary Treatment and Protection	18
	Separator	18
	Filters	19
	Wipers	19
	Boots	19
3 ·	**Distribution Systems**	21
	Reservoir	21
	Piping Systems	21
	Pitch	22
	Take Offs	23
4 ·	**General Schematic Symbols**	25
	Connected and Non-connected Lines	25
	Cylinders	26
	Flow Control	26
	Variable Symbol	27
	Compensated Flow Control Valves	27
	Check Valves	29
	Fluid Treatment	29
	Pressure Relief Valve	29
	Motors and Pumps	30
	Reservoir or Tank	31
	Accumulator	31
	Pilot Lines, Exhausts, and Enclosures	32
	Hydraulic or Hydraulic Arrowhead	34
5 ·	**Valve-Related Schematic Symbols**	35
	Self-Test #1	43
6 ·	**General Force Equation for Cylinders**	47
7 ·	**Hydraulic Fittings**	51
	Barbed Fittings	51
	Tubing Sizes	51
	Hose Sizes	52
	Quick Disconnects	53
	Hose Clamps	53
	Nipples	54

Contents

Straight Couplings	55
Reducers	55
Ferrule or Sleeve, and Nut	56
Manifold	57
Strain Relief	58
Swivel	59
Tee	59
Elbow	60
Valves	60
Plug	62
Runs and Branches	62
Male & Female	63
Union	64
Flow Control with Integral Check	64
Straight Connector	66
Pipe Thread	66
Summary	67
Self-Test #2	68

8 · The Ideal Gas Law and Solved Problems 71
  Solved Problem 73
  Solved Problem 74

Section Two

9 · The Hydraulic Side of Fluid Power 75
  Friction 75
  Viscosity 75
  Distance, Area, Volume 76
  Distance, Speed (Velocity), Acceleration 76
  Flow, Pressure 77
  Momentum (a 5th level tech. term) 77
  Force, Weight (6th level tech. terms) 77
  Energy, Torque (7th level tech. terms) 78
  Power, HP (8th level tech. terms) 78

10 · Hydraulic Terms and Concepts That Help Your Understanding 81
  Efficiency 81
  Pressure Differential 82
  Dissolved and Entrained Air 82

Cavitation	82
Actuator Speed, Pressure, Flow, Force, and Torque	83
Intensifiers	84
Fire Points	84
Fire Resistant Fluids	85
Hydraulic Cylinder Terminology	85
Hydraulic Accumulator Terms	90

11 · The ABCs of Hydraulic Relationships 93

12 · Hydraulic Motors 97

Positive Displacement Hydraulic Motors	97
Vane Motors	97
Gear Motors	98
Piston Motors	98
Variable Displacement Hydraulic Motors	98
Axial Piston Swashplate Motor	99
RPM	99
Torque	100
Horsepower	101
Freewheeling	101
Items & Terms	102
Self-Test #3	103

13 · Pressure Control Valves 105

Pressure Relief Valve	106
Sequence Valve	106
Brake Valve	107
Counterbalance Valve	108
Unloading Valve	108
Pressure Reducing Valve	109
Summary	110

14 · Proportional and Servo Valves 111

Six General Areas of Valve Control in Hydraulics	111
Electro-mechanical Control (Levers and Solenoids)	115
Hydrostatic Drive Control (Hydrostatic Drives)	115

Proportional Valve Control	116
Single Stage and/or Instrument Servo Control	117
Two Stage Servo Control	118
Three Stage Servo Control	119

15 · Discussion of Self-Test Answers 121
Test #1	121
Test #2	125
Test #3	128

16 · Exercises 131
Exercise 1	134
Exercise 2	135
Exercise 3	137
Exercise 4	139
Exercise 5	141
Exercise 6	143
Exercise 7	145
Exercise 8 (no build—from print only)	148
Exercise 9 (optional) (no build—from print only)	151
Exercise 10	153
Exercise 11	156
Exercise 12 (optional)	159
Exercise 13	163
Exercise 14	166
Exercise 15	169
Exercise 16	172

Preface

This book was developed to instruct people who want to troubleshoot hydraulic machinery and hydraulic circuits. The material in this book assumes no prior knowledge of hydraulics and could be used by anyone who has an interest in this particular area of fluid power. This book does not cover the rebuilding of hydraulic components.

In order to firmly plant the concepts of what is going on in hydraulics, this information has an orientation to a "hands-on" approach. The text uses some generalizations and other approximations. This book is directed at the hourly worker on the factory floor or out in the field.

The objective of this basic hydraulics course is to train the current or prospective maintenance mechanic or technician in the basic hydraulic building blocks to be used in troubleshooting. This is done in a two-step fashion. If you have had no previous experience with pneumatics or hydraulics, then start right at the beginning of section one. If you have had previous experience with troubleshooting pneumatics, then look over and review the first section of the text, do all of the exercises (1–8) in the first section, and then start in earnest on section two. This hydraulic text is oriented to what you will run into in the field. Technical specs and other values are given in the text and then followed by their {standard} and [metric] equivalent, as you should be able to work in either measuring system in this field.

Jay. F. Hooper
Salisbury, N.C.
December 2011

Introduction

This book is part of a course that prepares the student to display a working knowledge of hydraulic systems and to troubleshoot hydraulic problems. Upon successful completion of this course, the student will be able to:
- Read schematic prints
- Understand the components of hydraulic systems
- Recognize the names of hydraulic components and fittings
- Determine probable causes and solutions of problems
- Troubleshoot standard hydraulic circuits

Basic Hydraulics

Hydraulic Fundamentals 1

The first thing that we want to take a look at before jumping right into the components of a hydraulic system is some background. This information includes the various units and pressure scales used in hydraulics and some background on fluid flow in general.

The Four States of Matter

Matter as we know it is generally thought to consist of four states, in this order:
- a) solid
- b) liquid
- c) gas
- d) plasma

An example of a solid would be the table that you are using to write on. An example of a liquid would be the drink you had at the water fountain. The air that you breathe is an example of a gas. What about plasma? Look up. See the fluorescent lights up there? That eight footer [244 cm] has an electric arc inside of it that is about eight feet [244 cm] long. It exists inside of the glass tube due to the plasma state of matter. Let's examine water through three of these states. As a solid we know water as ice, while as a liquid we drink it all the time. If we boil water on the kitchen stove, we have all seen the "smoke" given off that we call steam or water vapor.

Change of State

As matter changes from one state to another, two very curious things happen. (1) The temperature remains the same. (2) It gives up or takes on heat. Let's illustrate:

Upon successful completion of this unit the student will be able to:

Recognize the terms associated with an understanding of pressure scales and other fundamental units used in hydraulics.

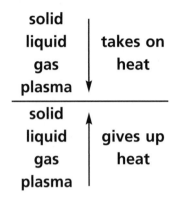

A few examples are in order. Take the example of water on a stove. We bring it to a boil at 100 degrees C {212 degrees F}. Now, we can boil and boil and boil, and eventually the pot will dry up. What happened? Well, all the liquid water changed into water vapor (change of state). What was the temperature at which this was done? 100 degrees C {212 degrees F}. Look at the arrow on the chart above. If you move from a liquid to a gas, matter takes on heat. Well, what about it? Did the water take on heat? You're absolutely right, it did. Ask the power company about it.

What about the ice in your iced tea? Well, this is what happens to me. (1) Put ice in a tall plastic tumbler. (2) Get a pot of room-temperature tea off the kitchen counter. (3) Pour tea into the tumbler. Crack, crack, blam (sometimes). The ice is going through thermal shock. I once had a piece of ice that made it a distance of 3 meters {10 feet} across the kitchen floor on the fly.

At any rate, the ice fairly quickly warms from say -10 degrees C {14 degrees F} to 0 degrees C {32 degrees F}. The tea temperature on the other hand is heading for 0 degrees C {32 degrees F} from say 32 degrees C {90 degrees F}. Is anybody's head spinning yet?

Now I have a nice cool drink at 0 degrees C {32 degrees F} as the ice continues changing state from a solid to a liquid. Of course, this keeps the tea cold because as you see from the chart on the previous pages, the ice must take on heat from the immediate environment (i.e. the tea and your hand, etc.) in order to change from a solid to a liquid at 0 degrees C {32 degrees F}.

But what about making that ice? If you go from a liquid to a solid, the chart says that matter gives up heat. Just feel the exhaust from your refrigerator as it is making a batch of ice. It heats up your house or apartment. Great for the wintertime, but we could do without it in the summertime.

So to summarize: as matter changes state, the temperature remains the same and depending on which way you travel up or down the list on the chart, it gives up or takes on heat.

STP

This abbreviation stands for standard temperature and pressure, and comes from the scientific and engineering areas. In the fluid power area of hydraulics this concept is usually called:

Normal Air

The hydraulics people are talking about air at 100,000 Pascals or 1 bar {14.7 PSIA}, 20 degrees C {68 degrees F}, and 36% relative humidity. The term "normal air" is used as a standard of comparison to what is called headspace air in a hydraulic tank.

Free Air

Free air is essentially the condition of the air that is around you, wherever you happen to be, as regards temperature, pressure, and relative humidity. Hence, the name "free." This air is in whatever condition that you find it in (i.e. hot and humid summer day vs. dry, crisp winter day). This is what is on top of the hydraulic fluid in the headspace of most tanks that are open to the atmosphere (i.e. non-pressurized tanks). It refers to the condition of the air around you or your equipment.

Relative Humidity

What we are talking about here is a concept analogous to a sponge and water. From our own experience we know that a sponge will absorb or take up just so much water, and will eventually reach a point where it becomes 100% saturated and will hold no more.

With the atmosphere, the air becomes the sponge and water vapor is analogous to the water. As the atmosphere absorbs water vapor, it will reach a point where it becomes 100% saturated with water vapor and will hold no more. The atmosphere, as is the case with the sponge, will hold various amounts of water vapor from nothing (0%) to complete saturation (100%). In dealing with the atmosphere, this is called relative humidity, and the measurement is given as a percent. Hence 50% R.H. (relative humidity) means that at this air temperature, the atmosphere is currently holding 50% (or one half) of the maximum amount of water vapor that it could hold.

This always seems to bring up the observation by someone that when it is raining, the relative humidity is 100%. Well, yes and no. The relative humidity where you are standing during a rainstorm (such as a porch just outside the screen door) many times will be in the high 90% range, but not 100%. If it were 100% R.H. where you were standing, everything inside the house would be coated in water.

It is generally 100% relative humidity at the altitude where the rain is coming from, say 1000 meters {3300 feet}. If you think that this is confusing, ask a pilot what a rainbow really looks like.

The important thing to remember about the concept of relative humidity is that it is a percent and that it is relative. The actual amount of water vapor that the air can hold varies with temperature. This along with the dew point is the nub of any problems with water in hydraulic systems. Example: At 80% R.H., 2.8 cu. meters {100 cu. ft.} of the atmosphere at STP will hold about 0.9 kg.* [8.8 newtons] or {2 lbs.} of water vapor at 35 degrees C {95 degrees F}. That same air at 21 degrees C {70 degrees F} will hold only about 0.45 kg.* [4.4 newtons] or {1 lb.} of water.

Where did the rest of this water go? It can't stay in the air as water vapor any longer when the air cools down. What it does is condense (changes state) on the inside of your tank if you are not careful.

Dew Point

Dew point is intimately tied up with the previous discussion on relative humidity. Dew point is the temperature to which you must lower the air in order to condense the water vapor (change state). In other words, for any given set of atmospheric conditions (free air), there will be a temperature to which if the air is lowered, the R.H. will become 100% and the water vapor changes to liquid water. This point is the dew point temperature (called the dew point).

This is why the lawn is often wet in the early morning. The general atmospheric pressure and the total amount of water vapor in the air doesn't change much through the night, but the temperature drops as the evening wears on. On many nights the temperature goes below the dew point, so water vapor starts condensing out of the atmosphere onto the grass (colder air lies close to the ground, that's why fog will generally form first in low lying areas).

---

* 0.45 kg. {1 lb.} of water vapor is equal to approximately 470 ml. {1 pint or 1/8 gallon} of water. Weight in the metric system is really in newtons NOT kilograms. So 1 lb. of water is really 4.4 newtons of water. Hey, it's not me, it's the shipping guys. Their documents have standard weight in pounds listed next to metric mass in kilograms. By convention they call both "the weight of the object," but it is really not true. Just remember, on the factory floor they use kilograms and say that 1 kg. = 2.2 lbs., but in real life and in calculations it is really newtons and 2.2 lbs. = 9.8 newtons.

FPS [MPS or mps or m/sec.]

This stands for feet per second [meters per second] and is basically a measurement used with various hydraulic pumps and motors to indicate how far the hydraulic fluid is moving each second. On the floor this is called how fast the fluid is moving.

Scales

Before jumping into the next topic (pressure scales that are utilized in hydraulic systems), we need to review the general rules that allow one to deal with scales in general.

To understand any scale all you need is the value of one unit and any two points on the scale.

Scales exist in order to allow us to make comparisons, and as such, we need as a minimum only three things to allow us to understand someone else's scale. We need to know the units of the scale and two points on the scale. That's it.

Absolute Pressure Scale

The units are Pascals, kilo pascals, bars or millibars, and pounds per square inch. In general usage they are usually abbreviated Pa, kPa, bar, mb, and PSI respectively. Two points are a perfect vacuum (outer space), which is 0 kPa absolute or 0 bar absolute {0 PSIA}, and sea level pressure (the beach), which is 101.3 kPa absolute or 1 bar absolute {14.7 PSIA}.

Gauge Pressure Scale

The units used are generally kilo Pascals [kPa], [bar], or pounds per square inch {PSI}. Two points are atmospheric pressure, which is 0 kPa gauge or 0 bar gauge {0 PSIG}, and typical system pressure, which is about 680 kPa gauge or about 68 bar gauge {about 1000 PSIG}.

What this means is that whatever the free air conditions happen to be (Denver or Los Angeles) as regards atmospheric pressure, the gauge is always zeroed up so that at atmospheric pressure it reads 0 (zero).

pressure gauge in PSI, kPa x 100, and bar units

Vacuum Pressure Scale

The units are millimeters of mercury abbreviated mm Hg [torr] {inches of mercury, "Hg}. Two points on the vacuum scale are a perfect vacuum, which is 101.31 kPa vacuum or [760 torr vacuum] {29.92 "Hg vacuum}, and sea level pressure, which is 0 kPa vacuum or [0 torr vacuum] {0 "Hg vacuum}. Now understand that on many standard vacuum gauges you will see 30 [102 torr] on the scale, but you will never reach it in practice. 29.92 [101.31 torr] is it.

Major Constituents of Air

Air is made up of approximately 78% nitrogen (N_2), 21% oxygen (O_2), and 1% other gases. Now, normally in a situation where you have an atmosphere of 21% pure oxygen, you would have a major explosion if you lit up a cigarette or turned on a light switch. However, the huge amount of nitrogen in the atmosphere mediates the danger considerably, so what we get instead is rust (a really, really slow burn). In hydraulics if you use oil as your fluid, there is a very real fire danger if you get a leak and the oil spills out. There is also an explosive danger in an oil air mix if the oil sprays out.

Compressibility

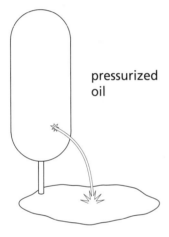

pressurized oil

Liquids have a compressibility of about 1 to 1, while gases on the other hand have a compressibility of about 1700 to 1. What does this mean to you? Well two things really. One, if you breach a receiver that contains a pressurized liquid (say a hydraulic oil system weight loaded or spring loaded accumulator), generally all that you get is a big mess on the floor as the liquid runs or shoots out. On the other hand, if you breach a receiver that contains a pressurized gas (say in a pneumatics system) of equal size and pressure to the hydraulic accumulator mentioned above, you would probably get an explosion. This explosion would probably injure or kill one or two people if they were standing next to your machine.

What makes the difference?

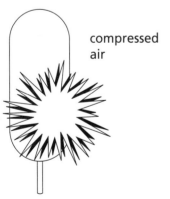

compressed air

The compressibility does. The compressed air in the breached container wants to expand to about two thousand times its original confined volume. We perceive and feel this as an explosion because it happens so rapidly. This is why receivers in pneumatic systems and tanks used for welding are positioned out of harm's way. Who wants some idiot on a forklift doing you and your buddies in?

The second thing compressibility means to the typical maintenance mechanic is this: If you have an accident and get caught in a pneumatic system (air), most of the time you can get your fingers, etc. out because a pneumatic cylinder at 90 PSI [620 kPa] has low pressure and some give (large compressibility factor). If you have an accident and get caught in a hydraulic system (liquid), most of the time you will NOT get your fingers, etc. out until the machine cycles again because a hydraulic cylinder at 500–5000 PSI [3447–34,474 kPa] has no give (no compressibility factor) and has higher pressures. A hydraulic system is not very forgiving to the human anatomy.

Understand this fact when working in the trades: breaching and getting your fingers caught are not things that occur that frequently in fluid power, but you should be aware of them.

Hydraulic Fluid Treatment or Conditioning 2

Primary Fluid Treatment

Primary fluid treatment is generally considered to be the conditioning of the fluid in a hydraulics system before the fluid enters the pressure line for the machine or process. Primary fluid treatment is generally located in and around the tank and pump or immediately adjacent to it. This contrasts with secondary fluid treatment, which is generally considered to be the conditioning or the protection of the fluid in a hydraulics system at or near the point of usage (the actuators). Of course, in real life sometimes these designations are flipped. In different trades the fluid conditioning around the actuators is considered primary and that around the tank secondary. Fluid treatment generally comes in a specific order (breather tube filters for headspace air, strainers, suction line filters, pressure line filters, return line filters, hydraulic coolers, tank baffles, reservoir sumps, etc.). A given system may not have all of them, but should have most of them.

Strainers

Strainers are generally of two types, either dry located in the throat of the tank's filler tube (also used as the air breather on many units) or wet located near the beginning of the suction line. The wet strainer filters may not be changed or cleaned very frequently, as many are generally hard to get to. A strainer is generally some type of fine wire mesh screen that is looking to capture any large (by hydraulic standards) particles that are floating around in the system.

Upon successful completion of this unit the student will be able to:

Recognize the makeup of the units of hydraulic fluid treatment or conditioning at the tank and at the actuators, and also the sequence of the units of hydraulic treatment or conditioning in a system.

Suction Line Filters

After the tank strainers, suction line filters are the second line of defense for the hydraulic system as a whole. These filters are sometimes also down in the tank, and are generally hard to get to.

Pressure Line Filters

Many times they do not look unlike the fluid filter in your car. Pressure line filters are the third line of defense for the hydraulic system as a whole after the suction line filter. These filters may need to be changed frequently, as many operations occur in dirty environments.

Return Line Filters

After the pressure line filter, the return line filter is the fourth line of defense for the hydraulic system as a whole. Again, these filters may need frequent changing, as many operations occur in rough environments.

Hydraulic Fluid Cooler

If your system has a hydraulic oil cooler, it would be located near your tank or reservoir. Generally, a lot of hydraulic coolers utilize air as the cooling medium. Other hydraulic systems use a fluid like water or even a water glycol mix as the cooling medium. Another name that you will hear for a hydraulic cooler is hydraulic oil cooler. They are often used interchangeably.

Hydraulic Fluid Heater

Yes, sometimes it is extremely cold. If your system has a hydraulic oil heater it would be located near your tank or reservoir, or inside your tank. Generally, most hydraulic heaters utilize electricity as the heating medium, although some systems sometimes use a fluid like water glycol in a heat exchanger. Another name that you will sometimes hear for a hydraulic heater is hydraulic oil heater.

Separation of Water

A filter separator uses a filter or a mechanical means to separate water. Examples of the filters that are used are a 3 micron filter in the breather tube to keep moisture from getting into the fluid via the headspace in the tank, or a plastic polymer filter that swells and traps moisture in a hydraulic line. Mechanical separation uses a sump or other mechanical means to get water droplets out of hydraulic fluid.

Separation of "Big" Dirt

The tank causes the hydraulic fluid coming through the inlet to make several turns around baffles and through a diffuser of some sort before going out the outlet (the suction line). The theory is that no foam will be generated and "big dirt" in the fluid will not be able to make those turns and will fall next to the inside wall of a baffle in the bottom of the tank. Then the particles will slide down and collect in a sump in the bottom of the tank. Here some of the captured dirt (and water) can be drained periodically if you are in a really dirty work environment. It usually depends on the original design of the system. Once a year or so when the oil is changed out completely, the tank should be opened and any sludge thoroughly cleaned out of the bottom of the tank.

Breather Adsorption Filter

In some special cases as the final step in primary fluid treatment, you might go after the water again with an adsorption filter. These filters generally use silica gel. This material is what comes in the little pouch inside of the box that your new stereo or camera comes in. It is put in that box to adsorb any moisture that might have entered the package and caused rust or corrosion of the merchandise before you buy it.

In hydraulic breather systems, you would generally have one of the adsorption filters operating while the other one is recharged or renewed. Once the silica gel adsorbs a full load of water, it can be reused again simply by passing warm dry air through the filter element and discharging this air into the atmosphere.

Remember, most hydraulic systems will not have all of these units of primary fluid treatment, but most of them will (or should) have most of these units.

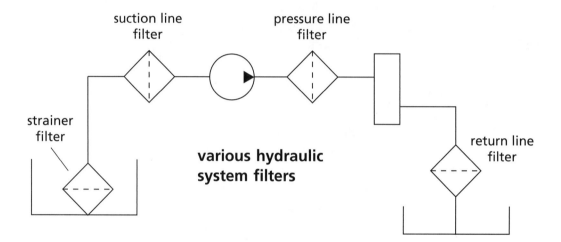

various hydraulic system filters

Micron

A micron is equal to 10^{-6} meter or one one-millionth (1/1,000,000) of a meter, a meter being a little bit larger than a yard. You will find micron abbreviated as μm, which is pronounced "mew, m." When you see that abbreviation, μ stands for the word micro, and m stands for the word meter. Hence, most people will say micro meter or micron when they see μm.

In a lot of hydraulics literature you will also see the abbreviation um "u, m." This also stands for micron and has come about because the standard computer keyboard does not have the Greek letter μ on it. There are about 25 microns in 1/1000 of an inch.

The filter that you will use in a particular situation is usually determined by the size of the particle that you want to stop.

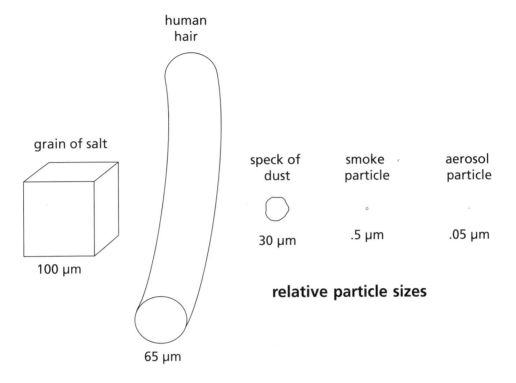

relative particle sizes

Filters are generally of two types, surface and depth. The surface filter uses a thin, closely packed element to do its work, and is generally a washable, reusable filter type. The depth filter uses a thicker, more loosely packed element, and is generally a throw-away filter type. Both types of filters, surface and depth, use a wide range of materials, such as metallic wire or ribbon, felt, cloth, paper, sintered metal powders, etc.

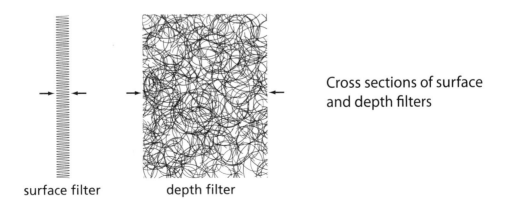

Cross sections of surface and depth filters

Filters also have what are called absolute and nominal ratings. That is to say, on the spec. sheet that comes with a new filter element, you may find an absolute and/or a nominal rating for the filter element. The ratings will be in microns. An absolute rating of forty microns, for example, would mean that this particular filter element, when used properly, will capture all particles larger than 40 microns. A nominal rating of 40 microns, for example, would mean that this particular filter element will capture some percent (say 95%, plus or minus two standard deviations) of all particles greater than 40 microns.

The nominal and absolute ratings are based on the bell shaped curve.

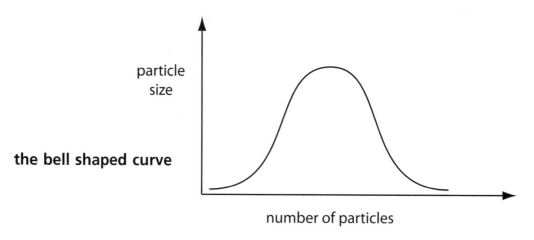

Secondary Treatment and Protection

When the oil gets close to the actuators, it often goes through a second level of oil treatment and protection (secondary treatment and protection). In most instances this consists of wipers and boots for your cylinders, etc., or even additional filtering.

Separator

A separator, if it exists, is generally part of a filter unit. This is of particular importance in outdoors hydraulic equipment usage. Any water collected will be drained off when necessary.

Filters

A filter is looking to capture particles on the order of microns so that the narrow valve passages in your hydraulic machine components do not get all gummed up and clog.

Wipers

These seals mechanically wipe the shiny metal areas of the cylinder rod in order to keep the inside components of the cylinder free of contamination from the job site. The cylinder rod needs a thin film of oil to continue working properly. The seals of a cylinder need the oil to remain flexible and to not dry out. The interior metal areas of the cylinder wall need oil for lubrication and to keep any rust, etc. from forming.

Boots

These barriers mechanically block off the outside shiny metal areas of the cylinder rod in order to keep the inside components and seals of the cylinder free of any heavy contamination from the job site.

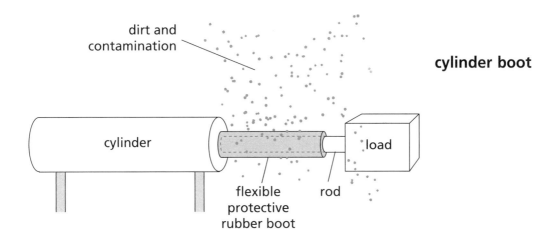

cylinder boot

2 · Hydraulic Fluid Treatment or Conditioning

Please understand that not all hydraulic systems have all of these units of secondary treatment and protection, but most of them will (or should) have most of these units, particularly "out in the weather" hydraulic units.

Distribution Systems 3

If installed properly, a hydraulic distribution system not only delivers oil (pressurized fluid) where you need it, but it also will not leak a lot.

Reservoir

The reservoir (tank) is utilized primarily as storage in a hydraulic system. This allows a short, rapid draw down of fluid without affecting the system's pressure to any great degree. Usually reservoirs (tanks) are very large compared to the size of the pump. They are protected from harm by being placed to the side of normal traffic patterns. This is to prevent the possibility of a messy spill by poor forklift driving habits or poor overhead crane habits, etc. These simple precautions and the quality of the tank are usually good enough so that you do not have a giant leakage problem.

Piping Systems

In hydraulic installations the main distribution system is typically a steel pipe, steel tubing, socket weld steel pipe, or a flexible hose, or some combination of the four.

Upon successful completion of this unit the student will be able to:

Recognize the appropriate makeup of a hydraulics distribution system and the application of subsystems and components.

Learning activities
- Lectures
- Readings
- Demonstrations
- Hands-on experience (Exercise #1)
- Evaluation and discussion of the lab exercise.

4000 PSI (275 bar) hydraulic hose

A hydraulic distribution system closes back in on itself in a loop (a closed system) and largely negates any large pressure drops in your machine's system due to distance from the pump.

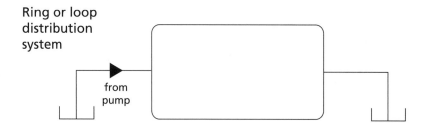

Ring or loop distribution system

While this loop actually exists in real life, on your machine the loop will not show up on your hydraulics print. Your print will look more like a multi-branch circuit. (This is an artifact from how prints are drawn.)

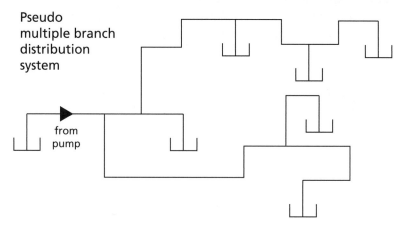

Pseudo multiple branch distribution system

The simple reason for this is that almost all hydraulic prints use more than one tank symbol on the print in order to cut down on the total number of lines drawn on the print.

Pitch

Generally speaking, the main distribution system should pitch up about one inch for every twenty feet of run. This is so you have a low spot in the beginning of your system where the tank is installed.

The worst thing that you could have would be a system where the main distribution lines are all level and there would be no system low point to properly install a tank in. In this type of system (which is, by the way, typical of many installations), the oil tends to lie in the low spots of various pipe segments throughout the machine, and has no place to drain to. This is a real pain if you need to change all of the oil out yearly or semi-annually during machine maintenance or overhaul.

Take Offs

At some point you need to tap into the main distribution system piping in order to get some oil (pressurized fluid) up or down to an actuator in the machine. The take off should be done with a tee of some sort in order to have a small flexible loop in the piping for absorbing vibration, etc. Some industries tee up with a flexible loop and some tee down.

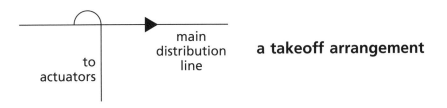

a takeoff arrangement

Please turn to page 134 and do Exercise #1 before continuing with the text.

General Schematic Symbols 4

Now we are going to look at the general schematic symbols that you will encounter on a typical hydraulics print. You would be using this print in a troubleshooting situation. We will be taking a building-block approach to this endeavor, and as such should touch on about 80% or more of the symbol information that you will need. For the other 20% or so of symbols that you will run into, just look them up or ask someone at your workplace about them. We will be concentrating on what most of you will be running into most of the time. You could spend hours and hours learning all the minor symbol aspects, but it would not be that helpful in your job situation.

Connected and Non-connected Lines

In drawing or interpreting lines in hydraulics or any discipline you are in (be it electrics, electronics, or whatever), there is a very good unambiguous way of telling someone whether or not a line is connected or not. Use a dot at the intersection of connected lines:

Use a half circle where lines cross but are not connected on the print:

Upon successful completion of this unit the student will be able to:

(1) Recognize the general schematic symbols associated with hydraulics, and
(2) determine the relationship between pressure and flow.

Learning activities

- Lectures
- Readings
- Demonstrations
- Hands-on experience (Exercise #2)
- Evaluation and discussion of the lab exercise.

If you do these two very simple things, then everybody will understand your meaning. What does the following drawing mean?

Who knows? Why guess? Use the dot to show connection and the half circle to show lines crossing but not connected.

Cylinders

In hydraulics there is one main type of cylinder, the double acting one. Single acting hydraulic cylinders and other types of hydraulic cylinders also exist but they are in the minority.

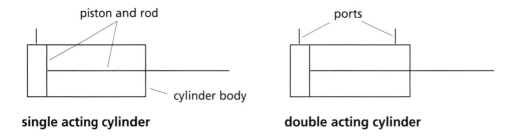

single acting cylinder   **double acting cylinder**

The schematic symbol for these actuators shows the cylinder body, the piston and rod, and the number of ports each cylinder has. The double acting cylinder uses oil to extend and retract, while the single acting cylinder uses oil to extend, and usually gravity from a supported load to retract or vise versa. That's why the single acting cylinder has only one port.

Flow Control

A flow control valve is a device that is placed somewhere in the system in order to regulate the flow to or

4 · Distribution Systems

from an actuator (usually). This could be thought of as a speed control. The symbols would be as follows:

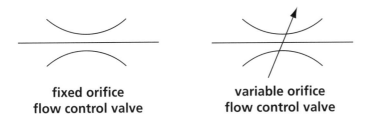

The difference between these two devices is that the fixed orifice flow control valve is not adjustable, while the variable orifice flow control valve is.

Variable Symbol

No matter what discipline you are working with in your job, if you find a leaning arrow superimposed over another schematic symbol, then the word "variable" will be in the description somewhere.

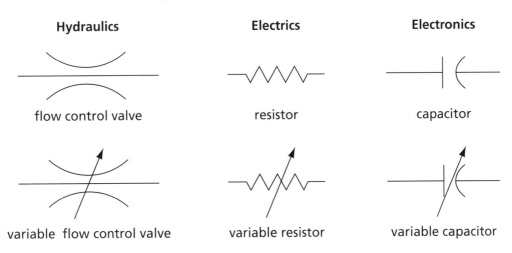

Compensated Flow Control Valves

Within the hydraulics world, a situation can occur in which over time, the fluid that you use heats up and becomes "thinner." What this means with a standard flow control valve is this: over time more fluid flows through the valve at any given setting, which means that the affected actuator will go faster (usually up to 15% or so). To keep that from happening, you can use pressure or temperature compensated flow control valves. These valves will automatically compensate for the "thinner" hydraulic fluid. That way your actuator will go at the same set speed regardless of the oil temperature and the "thinness" of your hydraulic fluid.

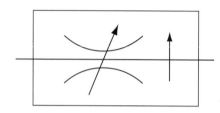 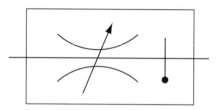

pressure compensated flow control valve

temperature compensated flow control valve

temperature and pressure compensated flow control valve symbols

the outside physical package of a pressure compensated valve
(It looks the same as a regular flow control valve, just bigger.)

Check Valves

A check valve is a device that allows flow in one direction and no flow in the opposite direction. In hydraulics this is called free flow and no flow.

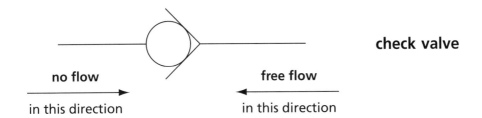

check valve

no flow in this direction

free flow in this direction

Fluid Treatment

The general symbol for fluid treatment in hydraulics is the diamond shape. This is true whether you're talking about a filter to get rid of dirt, a filter separator to get rid of water, or a cooler to hold down the temperature of your working fluid.

diamond shape

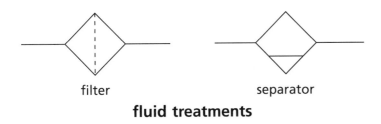

filter   separator

fluid treatments

At this point in time, all that you need to know is that when you are looking at a hydraulics print and you see a diamond-shaped symbol on it, the symbol represents some type of fluid treatment. Please note that wipers and boots are considered part of treatment, but as a rule do not show up on schematic prints.

Pressure Relief Valve

A pressure relief valve in hydraulics is used to regulate the oil pressure of the system. Generally, this relief valve is adjusted infrequently, or not at all, once the initial setup has been done. The general symbol for a pressure control valve is a square. A pressure relief valve is a subset of the pressure control valve square family of symbols.

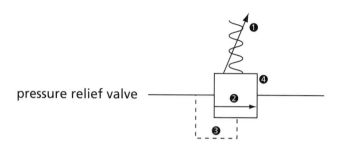

pressure relief valve

The little doodads on the symbol are: (1) the leaning arrow superimposed on the spring—the word "variable," (2) the internal arrow not lining up with the entrance and exit port of the valve—the word "normally closed," (3) the dotted line—internally pilot operated, and (4) the square—the word "pressure control valve." So, you officially have a normally closed, internally pilot operated, variably adjustable pressure control valve. In the field of hydraulics, on the floor, in catalogs, and on prints this is called a pressure relief valve. You just need to able to recognize this symbol as the pressure relief valve on a hydraulics print and to be able to locate it when you are troubleshooting.

Motors and Pumps

In general, in all the fluid power disciplines you will be dealing with, a device that rotates is associated schematically with the circle. Of course, not all circle symbols represent rotation. Here is an electrical shunt coil, for example:

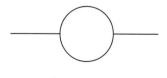

shunt coil

On the hydraulics print, a hydraulic motor is generally represented like this:

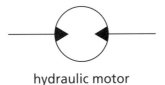

hydraulic motor

And pumps are generally represented in this manner:

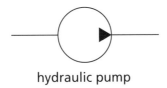

hydraulic pump

You need to be able to recognize the basic differences between electric motors, hydraulic motors, and pumps. You also need to be able to distinguish between the direction of rotation and the direction of flow.

Reservoir or Tank

The symbol for a reservoir (tank) of oil is:

tank

Accumulator

An accumulator is a reservoir of pressurized oil that is very small most of the time, say, on the order of twenty cubic inches. This is comparable to say, ten cubic feet or more for a receiver used in a compressed air system. Accumulators are generally used on production machines to provide an extra boost to an actuator or to provide some surge protection.

accumulator

accumulators

You will notice that accumulators use (1) the same general schematic symbol as a compressed air receiver (indicating some type of reservoir), (2) a horizontal line indicating the piston or diaphragm that separates the two compartments of an accumulator, and (3) an indicator of whether the accumulator is spring loaded, gas charged, or weight loaded.

spring loaded accumulator   gas charged accumulator   weight loaded accumulator

The gas charged accumulator usually uses dry nitrogen as the gas for safety reasons. If there is an accidental breach of the dry nitrogen charged accumulator, there will be no secondary explosion. If, on the other hand, compressed air were used in gas charged accumulators and a breach occurred, you would have an oil and air mixture, which is highly explosive in many instances.

As you may recall from your safety training, an empty 55 gallon drum with just a tad of material in the bottom is much more prone to explosion then a full drum of the same material. This is due to the vapor and oil mixture that exists inside of the "empty" drum.

Pilot Lines, Exhausts, and Enclosures

Troubleshooting in hydraulics exposes the mechanic to many prints drawn by many different draftsmen. The prints may differ somewhat in the details, but the basic building blocks are the same. In general, dotted lines on hydraulic prints fall into four categories. If the dotted line closes back in on itself, it stands for an enclosure. This means that you could go to the machine and find that whatever devices were inside of the dotted line on the print, the real world components would

4 · Distribution Systems

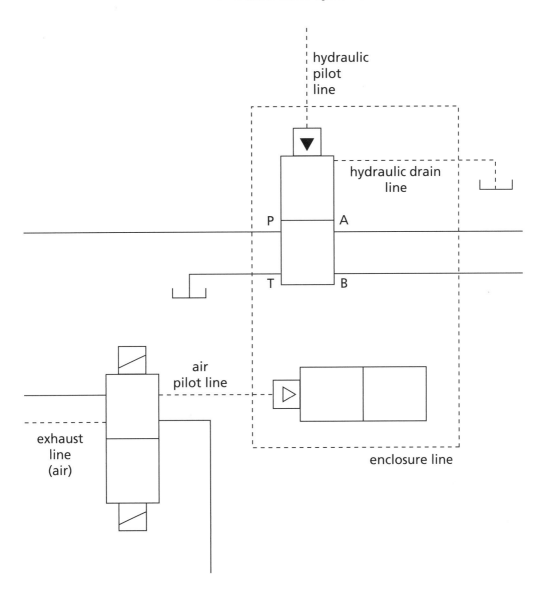

be inside of a separate enclosure or on a separate sub panel on the machine.

If a dotted line on a print goes from a device and just ends on the print (i.e., does not connect to anything), then it indicates a place in your circuit where the air from an air pilot line that operates a hydraulic directional control valve exhausts to the atmosphere.

A dotted line that goes from one part of the circuit to another part of the circuit will be a pilot line (i.e., there is a connection on both ends of the dotted line). Pilot lines are used to trigger components in a hydraulics circuit and do not handle the main oil flow of the circuit.

A dotted line that goes from one part of the circuit to a tank symbol will be a drain line. Drain lines are used to release any trapped fluid within components in a hydraulics circuit, and do not handle the main oil flow or pressures of the working circuit.

Hydraulic or Hydraulic Arrowhead

If you run into a small triangle

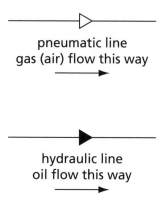

on a hydraulics print, it indicates flow. If the triangle is filled in or solid, then it indicates liquid flow (hydraulic oil). If the triangle is open, it indicates gas flow (air).

Please turn to page 135 and do Exercise #2 before continuing with the text.

Valve-Related Schematic Symbols 5

By this point in time, you have been exposed to many of the basic schematic building blocks of a hydraulic circuit. There remain several more major categories to look at, and directional control valves are what we will look at next. (Pressure control valves will be covered in a later chapter.) Directional control valves are the components that act like switches in a hydraulic circuit. The directional control valve is recognized by its basic shape of a rectangle.

The directional control valve rectangle

The rectangle is further divided into squares. These squares indicate how many positions (states) the directional control valve has.

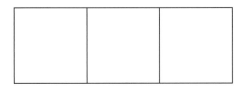

**3 position
directional control valve**

**2 position
directional control valve**

Upon successful completion of this unit the student will be able to:

Recognize directional control valve related schematic symbols and construct a circuit using general schematic symbols.

Learning activities

- Lectures
- Readings
- Demonstrations
- Hands-on experience (Exercises 3 and 4)
- Evaluation and discussion of the lab experience.
- Written Self-Test #1

Three position and two position directional control valves are the most common. You will not find a one position directional control valve, and your chances of finding a four position or more directional control valve are small.

The next thing to determine is the number of ways the directional control valve has. The general rule is this: look at the directional control valve symbol. Pick out the position that has the most arrows in it. Multiply by two.

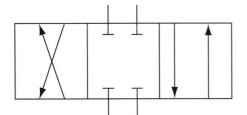

4 way, 3 position, closed center, directional control valve

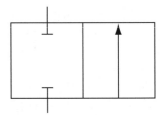

2 way, 2 position directional control valve

The term "way" came about in this manner. The arrow indicates which way (regular English usage) the oil flows through the valve in that particular circuit. The oil could just as easily flow through the valve in the other direction, if it is hooked up that way. Hence, for any given arrow in a particular position, the oil could flow through the valve either way depending on how the circuit is hooked up.

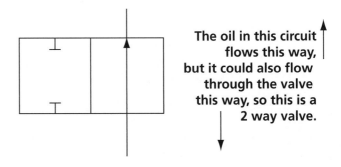

The oil in this circuit flows this way, but it could also flow through the valve this way, so this is a 2 way valve.

There is one major exception to this general rule. You will need to memorize it. It is the three way valve.

5 · Valve-Related Schematic Symbols

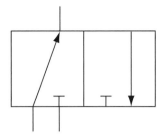

3 way, 2 position directional control valve

The name three way comes from the old technique of determining the number of ways of a directional control valve by counting the number of ports. Hence, this is a three way valve:

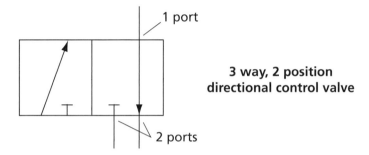

3 way, 2 position directional control valve

It is not a two way valve.

Remember, memorize what a three way valve looks like and determine the ways of all other directional control valves by applying the general rule: pick out the position of the valve that has the most arrows in it and then multiply by two.

The number of ports that a directional control valve has is determined by looking at any position and then seeing how many "things" touch the outside perimeter of the square. So, if you are looking at the following valve symbol for the air pilot line of your big hydraulic valve on a print,

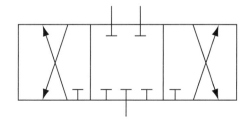

3 position, 4 way, 5 ported, directional control valve

and on the machine that you are troubleshooting, you are looking at a directional control valve for an air pilot line, but with two ports on one side of the valve body and one port on the other side of the valve body, you are in big trouble. Do you know why?

These little tee shaped items (T) that you have been seeing in these valve symbols are internal stops. What this means is, that a particular valve port, on a particular position of the directional control valve, is blocked internally. Hence, the name internal stop. If the port of a directional control valve is blocked by a plug (externally) it will show up as an X on the print on one of the valve positions (usually the rest position).

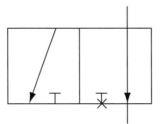

2 position, 3 way directional control valve set up as a 2 position, 2 way valve

The larger hydraulic directional control valves need to be triggered in some manner to switch oil flow on and off to the main circuit components. The three main ways to do this are by using an electrical solenoid, using a manual method (either hand or mechanical), or by using pilot pressure (either hydraulic or pneumatic). This is shown on the print by attaching the appropriate symbol onto the end of the directional control valve rectangle.

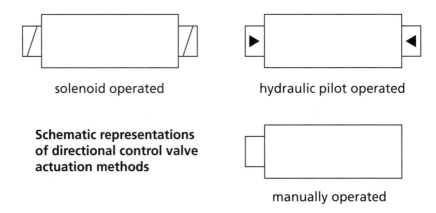

Schematic representations of directional control valve actuation methods

If the directional control valve has a spring symbol associated with it, then the following applies: you would have a spring return valve with one spring symbol, and you would have a spring centered valve with two spring symbols.

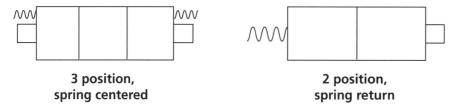

**3 position,
spring centered**

**2 position,
spring return**

Two position manual directional control valves (in hydraulics) are generally spring return, while three position directional control valves are generally spring centered (i.e., in the deactivated state, the spring on either end of the valve body moves the spool of the valve into the center position).

The solenoid operated directional control valve symbols are pretty straightforward as far as interpretation is concerned. The pilot operated directional control valve symbols have several major categories that you should be aware of.

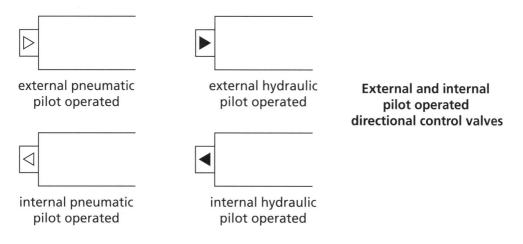

external pneumatic
pilot operated

external hydraulic
pilot operated

**External and internal
pilot operated
directional control valves**

internal pneumatic
pilot operated

internal hydraulic
pilot operated

If a valve has an external pilot, it means that the pilot pressure is coming into the valve through a separate line (pilot line) from another part of the circuit. If, on the other hand, a valve has an internal pilot, then the pilot pressure is coming into the valve through a machined passageway in the valve itself (from the pressure port).

Another aspect that the mechanic or technician has to deal with is the combination of several of these valve operated symbols on the end of just one directional control valve rectangle. This occurs quite frequently in real life. The general rules of interpretation are as follows:

If the symbols are stacked this way

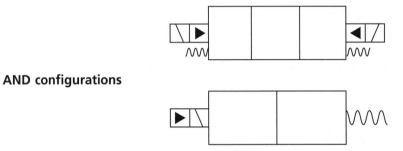

AND configurations

on the end of the directional control valve rectangle, then it is an AND situation. If the symbols are stacked this way

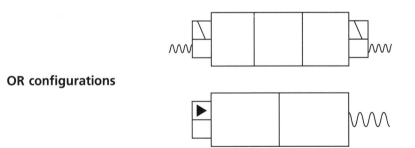

OR configurations

on the end of the directional control valve rectangle, then it is an OR situation.

And last but not least is the center condition of the directional control valves that you use in hydraulics. The four ports are generally labeled PTAB on a print, and on the valve itself in many situations. This stands for pressure port (P), tank port (T), actuator port (A), and actuator port (B). P comes from the pump. T goes to the tank. A and B go to your actuator. If you reverse the A and B ports your actuator runs backwards. The main types of center conditions are:

(1) tandem center

tandem center directional control valve

This is used to unload the pump in the rest position (P connects to T), but it keeps your actuator blocked and locked in the rest position.

(2) closed center

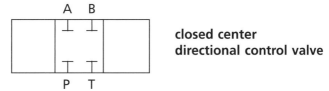

closed center directional control valve

This is used to block the pump and tank in the rest position, and it keeps your actuator blocked and locked in the rest position.

(3) open center

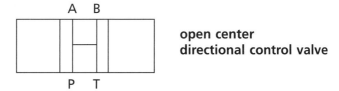

open center directional control valve

This is used to unload the pump in the rest position, since P is connected to T, and it keeps your actuator loosey goosey in the rest position (PTAB are all interconnected in an H pattern).

(4) float center

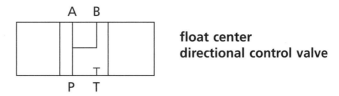

float center directional control valve

This is used to have the pump keep pressure on your actuator ports in the rest position (P connects to A and B, while T is blocked), but it keeps your actuator flexible in the rest position so that it can move a bit. This is sometimes called the snow plow circuit.

Sometimes you might have a combination like this:

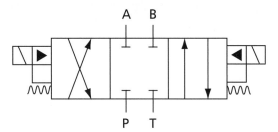

**three position, four way, closed center,
solenoid AND hydraulic pilot, OR manually
operated, spring centered directional control valve**

Please notice the placement of the commas in the above illustration's description. They become very critical. For example, this description would be wrong: a three position, four way, closed center, solenoid, AND hydraulic pilot or manually operated, spring centered directional control valve. Do you see why? The above valve description would look like this on a print:

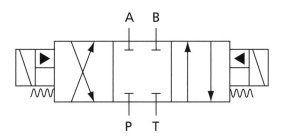

This is not the same valve as the illustration at the top of the page. It will not work in the same way.

Please turn to page 137 and do Exercise #3 and Exercise #4 before continuing with the text.

Next, you will take a self-test of the material covered so far. A discussion of the answers is located in the back of the book. Do not start the test until you are ready to begin. Answer all of the test questions before looking up any of the answers.

5 · Valve-Related Schematic Symbols

Self-Test #1

1) List the units of hydraulic fluid treatment in the order that you would normally find them. Remember, not all systems will have all of them. -30-

2) List the type(s) of distribution system piping in hydraulics, and in pneumatics. -10-

3) List the four states of matter in order. -5-

4) Describe what happens as matter goes from one state to another. -5-

5) What is the compressibility of oil vs. air? -5-

6) Define free air. -5-

7) Define STP as regards hydraulic tank headspace air. -5-

8) What is relative humidity? -5-

9) List or sketch several mechanical solutions to the contamination problem in hydraulic machines and systems. -10-

10) What is the absolute pressure scale? -10-

11) Identify the following symbols: -125-

A. 25 points

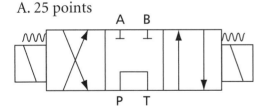

B. 25 points

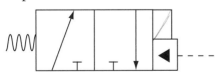

C. 5 points

D. 5 points

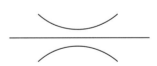

E. 5 points

F. 5 points

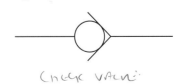

Check valve

5 · Valve-Related Schematic Symbols

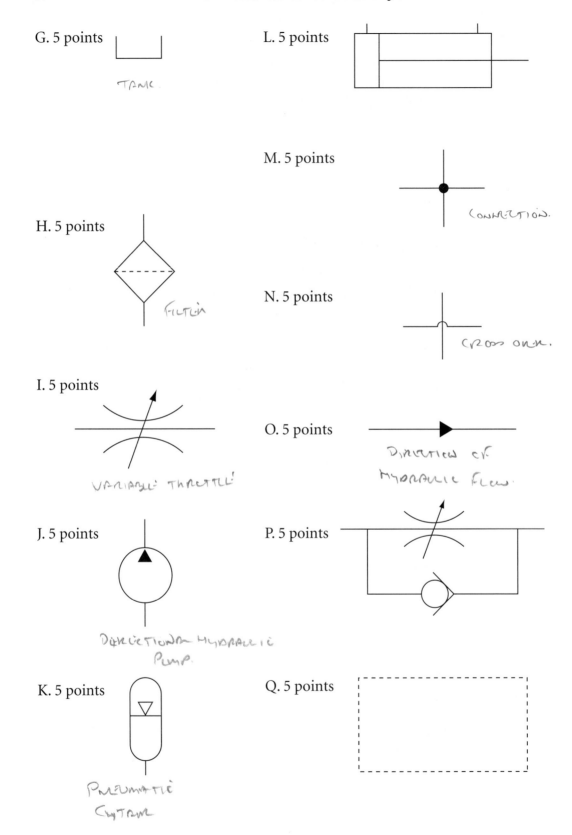

12) What is dew point? -5-

13) What is the gauge pressure scale? -10-

14) What does hydraulic fluid consist of? -5-

15) What is the vacuum pressure scale? -10-

16) What is the direction of the metered flow? -5-

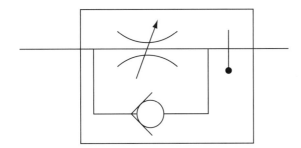

17) What is the direction of free flow? -5-

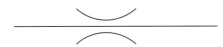

18) What does STP stand for? -5- STANDARD TEMPERATURE + PRESSURE

19) What is this? -5-

Bi Directional Hydraulic Pump.

265 possible points

General Force Equation for Cylinders 6

In talking about a cylinder, you cannot use the terms this end or that end, or the front end or the back end. These are all relative terms and would change depending upon the particular mounting orientation of the cylinder. Hence, in this text we will use the words cap end and rod end to denote the ends of the cylinder. These terms seem to conjure up the easiest visual image. Another term for the cap end is the blind end. Other terms for the rod end include annular end and head end.

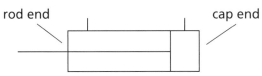

double acting cylinder nomenclature

The advanced hydraulic troubleshooter needs to be able to solve a handful of equations, but the average maintenance mechanic in basic hydraulics really only needs to be able to solve two equations:

$$F = PA$$

where F = the force in pounds [newtons]
P = the input pressure in PSI [Pa]
A = the area of the piston face in sq. inches [sq. meters]

and

$$A = \pi r^2 \text{ (pi r squared)}$$

where A = the area of the piston face in square inches [sq. meters]
π = 3.1416 ...
r = the radius of the piston face, or 1/2 of the bore of the cylinder

Upon successful completion of this unit the student will be able to:

- Recognize the relationship between force, pressure, and area in hydraulic circuits.
- Solve written problems related to force pressure and area.
- Measure and calculate the relationship between force, pressure, and area.

Learning activities

- Lectures
- Readings
- Demonstrations
- Hands-on experience (Exercise 5)
- Evaluation and discussion of the lab experience.

6 · General Force Equation for Cylinders

Understand that in your usual endeavors, you will not be called upon to do this all the time, but it does come up occasionally. A typical problem might look something like this:

We need about 18,000 lbs. [about 80,000 newtons] of force in this particular application. What size hydraulic cylinder should we order? The stroke will be about fourteen inches [350 mm].

$$F = PA$$

The system pressure on this machine is 900 PSIG [6205 kPa] or [6,205,000 Pascal] minimum.

$\quad$ F = (900 PSI) A $\qquad$ [F = (6,205,000 Pa) A]

The force that we need is 18,000 lbs. [80,068 newtons] or [What some of you would say is 8160 kg. but in reality is 80,068 newtons. (8160 kg. times g, the acceleration of gravity) or (8160 kg. times 9.81 m/sec. squared).]

$\quad$ (18000) = (900) A $\qquad$ [80,068 = 6,205,000 A]

Solving for A:

$\quad$ 900A = 18000 $\qquad$ [6,205,000A = 80,068]
$\quad$ A = 20 sq. inches $\qquad$ [A = .013 sq. m]

So we now know that we need a piston face of 20 square inches [.013 sq. m]. Cylinders are ordered by the bore. We must solve for r in $A = \pi r^2$, as this is equal to one half of the bore size (the inside diameter) of the cylinder.

$$A = \pi r^2$$

We know that A = 20 sq. inches [.013 sq. m], and pi = 3.1416

Substituting:

$\quad$ (20) = 3.1416 r^2 $\qquad$ [.013 = 3.1416 r^2]

Hence, solving for r^2 (r squared) we have:

$\quad$ 3.1416 r^2 = 20 $\qquad$ [3.1416 r^2 = .013]
$\quad$ r^2 = 6.37 $\qquad$ [r^2 = .0041 m]

Taking the square root of both sides we get:

$\sqrt{r^2} = \sqrt{6.37}$ $\quad\quad\quad [\sqrt{r^2} = \sqrt{.0041}\ ]$

r = 2.52 inches $\quad\quad\quad$ [r = .064 m or 64 mm]

d = 2r

d = 5.04 inches $\quad\quad\quad$ [d = 128 mm]

Since we need a minimum of 18,000 lbs. [about 80,000 newtons] of force, we would probably order a cylinder with a 6 inch [150 mm] bore. It is relatively easy to insert a pressure relief valve in front of this cylinder to cut down a little on the pressure and, as a result, the force. It is expensive to try to boost the force of a cylinder if you are already at the normal hydraulic system line pressure and are falling short by 5 or so percent. Don't cut things too close.

Please turn to page 141 and do Exercise #5 before continuing with the text.

Hydraulic Fittings 7

This section of the text speaks to the work situation that the average troubleshooting mechanic finds himself in. Generally, you are working with hydraulic systems at 300 PSIG [2068 kPa] to 2000 PSIG [13,790 kPa] and will either be a gofer (a helper) or will need to be able to order fittings (say, over the phone) if you are in charge. At any rate, you need to know the names of some of the most common hydraulic fittings that you will be using.

Some of these names will look a little strange to those of you who are familiar with some terms from plumbing or general pipeworking. To those of you who have no background in piping, no problem.

Barbed Fittings

If you have a situation where a fitting is attached to a flexible hose by being inserted into the hose and then secured by a hose clamp, then the word barbed will show up somewhere in your description of that fitting. But shame on you. These are NOT for hydraulic oil system use. These are generally for pressurized systems under 1000 kPa {145 PSI}, such as compressed air or low pressurized water lines.

Tubing Sizes

Tubing sizes are generally given by measuring the OD (outside diameter) of the tubing involved. This contrasts with most pipes that you will work with. The sizes in pipework are generally given by measuring the ID (nominal inside diameter) of the piping involved until you get up to 14 inches in diameter. Then pipes are generally measured by their OD. In hydraulics, two

Upon successful completion of this unit the student will be able to:

- Identify hydraulic fittings.
- Apply the meter in and meter out process to hydraulic actuators.
- Construct a hydraulic circuit using standard schematic symbols.

Learning activities

- Lectures
- Readings
- Demonstrations
- Hands-on experience (Exercises 6 and 7)
- Evaluation and discussion of the lab experience.
- Written Self-Test 2

common tubing sizes are 3/8" [10 DN] and 1/2" [15 DN]. You will use these two sizes a lot of the time.

tubing connectors

Hose Sizes

Hose sizes are generally given by measuring the ID (inside diameter) of the hose involved. This contrasts with most tubing that you will work with. The sizes in tubing are generally given by measuring the OD (nominal outside diameter) of the tubing involved. In hydraulics two of the most common hose sizes are 1/2" [15 DN] and 5/8" [18 DN]. You will use these two sizes a lot of the time. Most hoses are built by swaging (crimping) the ends on.

hydraulic hose ends with flange end elbows for crimping

Quick Disconnects

In some situations you may have a device that runs off pressurized fluid and that needs to be moved temporarily to another work station, somewhere in the plant. You may also have the need to use a test point for troubleshooting your hydraulic equipment. In order to do this safely and with a minimum of system oil loss, use a quick disconnect.

By looking down the inside of a quick disconnect, you can readily see that at the component level that there is a valve in the disconnect. You should not see any light through it.

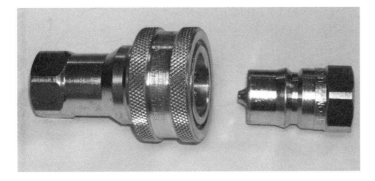

double shutoff quick disconnect (for pipe thread use)

The most popular quick disconnect in hydraulics, hands down, is the double shutoff quick disconnect. The double shutoff quick disconnect has a valve in both portions of the disconnect so that any leakage of oil is practically nil at the time of separation. This type of disconnect will be much more prevalent in hydraulics than in pneumatics. In pneumatics they use a lot of single shutoff quick disconnects. In hydraulics the pressurized oil system is usually shut off completely, or a portion of the system may be turned off with a valve when a temporary move is to occur somewhere in the hydraulic system.

Hose Clamps

Hose clamps go by a few other names: radiator hose clamps, aviation clamps, etc. They are used to secure a flexible hose to a barbed fitting of some sort. They are NOT used in hydraulic oil systems. They are generally

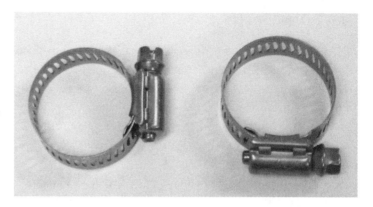

hose clamps

used at 1000 kPa {145 PSI} and below for compressed air, and for potable and low pressure water systems at around 300 kPa {about 45 PSI}.

Hose clamps are usually ordered by the nominal outside diameter of the hose that they will encircle. You DO NOT want to use hose clamps in regular hydraulic oil systems.

Nipples

Nipples come from piping and follow the same terminology. In hydraulics, most of them are made from brass, steel, or stainless steel. They would be ordered by (1) stating the nominal pipe ID, (2) the length of the nipple in inches [mm], and (3) the material that the nipple is made of. Say: ten, 1/4 inch [8 DN] x 3 inch [75 mm] brass nipples, or five, 3/8 inch [10 DN] x 4 inch [102 mm] steel nipples. Understand that if you actually measured the inside diameter of the nipples ordered above, you would not get 1/4 inch [8 DN] and 3/8 inch [10 DN] respectively. The nominal pipe ID is based on the approximate inside diameter of an equivalent black pipe nipple.

brass nipples (not to scale)

The bottom line is: regardless of the actual ID of the nipple, the pipe threads on the outside of the nipple will generally be the same no matter the composition or the wall thickness of the nipple. A 1/4 inch [8 DN] steel nipple will match up with a 1/4 inch [8 DN] female pipe thread on a brass hydraulic piping component just fine.

7 · Hydraulic Fittings

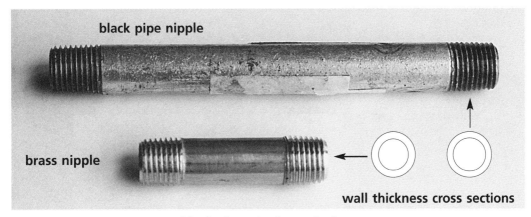

black pipe nipple equivalent

Straight Couplings

These fittings have two female pipe threads and are ordered by the pipe thread size and composition, such as: eight 1/4 inch [8 DN] brass, straight couplings. Again, don't be fooled by the wall thickness. Look at the pipe thread.

brass couplings

Reducers

The reducers that you will run into are generally of two types, the bell reducer and the reducing bushing. It is unusual to find a bell reducer made of bronze or brass in the typical hydraulic system, although bell reducers are quite common in the black pipe portions of

reducing bushings

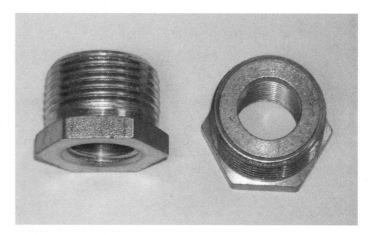

a factory compressed air system. Reducers are used to go from one pipe ID to another, say from 3/4 inch [20 DN] to 3/8 inch [10 DN]. The reducer bushings are used in combination with straight couplings or pipe nipples.

Ferrule or Sleeve, and Nut

In order to hold hydraulic tubing securely to a circuit segment or to a tube, or to a pipe adaptor, a compression type of system is utilized. This system consists of (1) a fitting that the tubing fits into, (2) a ferrule or sleeve that fits over the tubing and sits down against the fitting, and (3) a nut that fits over the hose, screws onto the fitting, and jams the ferrule or sleeve against the tubing.

compression fitting nuts

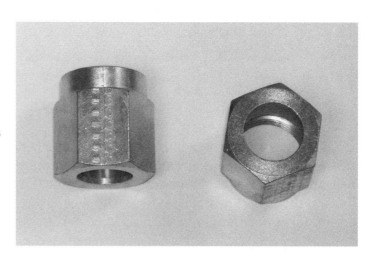

compression fittings

This gives a firm compression fit for a hydraulic system and is by far the most popular way to go. Who wants to use a jury rig with barbed fittings and hose clamps and have that blow up after the pressure passes through 200 PSI [1380 kPa]?

Manifold

In hydraulics it is often necessary to supply several parts of your system or your circuit components with oil from one point or junction. In general, if this component has more than three ports it is called a manifold. The actual manifold may be machined out of a

manual valves manifold

rectangular solid, a casting, or even put together from a large diameter square structural member that has been capped at both ends and drilled and tapped in the middle. Whatever the appearance on the machine itself, the manifold performs the same function in actual use.

Strain Relief

In many situations a worker needs a small amount of compressed air in order to run a small tool (such as an air ratchet) or to blow down his work area. The tool or nozzle is located at the end of a flexible air hose for ease of use. The other end of this hose, where it attaches to the air line at the wall, will very often kink and cut off the air flow.

To prevent this from happening, some type of strain relief is usually installed to prevent kinking. This is one type:

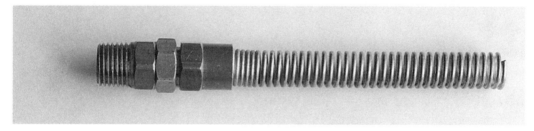

blow off hose, strain relief

Strain reliefs of this type are also occasionally used in hydraulics, but the use of a flexible cable tray unit is much more prevalent in this area of fluid power movement.

flexible cable tray

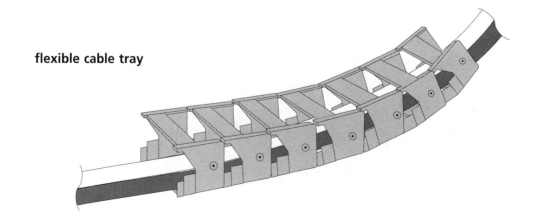

Swivel

A swivel is sometimes added (very sparingly due to the pressures involved) to a hydraulic fitting in order to allow additional movement of the flexible line to a greater degree without kinking. Of course, no oil will leak out of the swivel area while it is stationary or moving. If a component has a swivel feature, the phrase "with swivel" is usually put on the end of the general description, such as: twelve male run, quick connect elbows, with swivel.

elbows with swivel

Tee

This term refers to a junction that has three ports or connections, and is in the shape of the letter T.

8 DN (1/4 inch) tees

Elbow

Here we have a component that allows the hydraulic line to make a right angle, or a 45-degree turn. The name comes from the configuration of the human elbow when it is bent.

45 degree elbows

Elbows are not shown on hydraulic prints. The right angle turns that you see on a hydraulic print are part of the drawn hydraulic lines on the print, but they do not translate into elbows out on the machine. Elbows are used to allow the actual builder of a unit some flexibility in running the hydraulic lines where he or she needs to.

Valves

In hydraulics you will generally be dealing with three types of valves in the strict sense of the word (meaning to turn off or to turn on flow, completely or partially). Two of these come to us from piping.

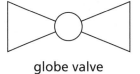

globe valve

gate valve

7 · Hydraulic Fittings

Now would be a good time to stop and look at a sample of each one. (Go check out a good hardware store, also known as big box heaven.)

The third type of valve is the needle valve. In hydraulics the needle valve is generally inside of components that schematically have the leaning arrow. You remember the leaning arrow, don't you?

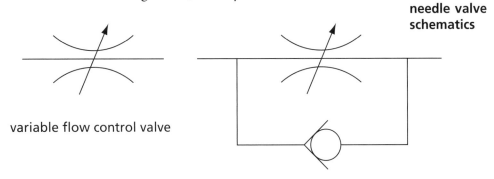

needle valve schematics

variable flow control valve

variable flow control valve with integral check

Needle valves allow for a fine adjustment of flow, and you will run into them quite frequently in hydraulics and also in pneumatics.

variable flow control with needle valve visible

Plug

Plugs are used to "stop up" unwanted or unused ports on a hydraulic device.

hydraulic plugs with o-rings

Runs and Branches

Sounds like a muddy creek situation. Run and branch refer to tees, not trees. The tee has three possible connections. The name that you use for a particular type of tee keys off of which connection has pipe thread or the main connection on it. Sound confusing? Well, it is easy to sort out, but first do the following:

Please turn to page 143 and do Exercise #6 before continuing with the text.

In hydraulics all tees have a branch and a run.

The name of a particular tee keys off of the connection (port) that has pipe thread on it or the main connection on it. Most compression fitting tees have pipe thread on one connection and nuts and ferrules on the other two, or have a nut and ferrule on one connection and machine thread on the other two.

37 degree flare fitting, male run tee

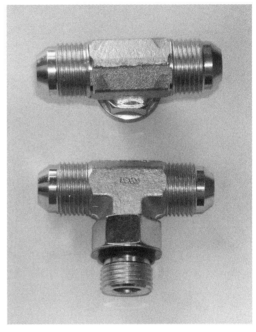

37 degree flare fitting, male branch tee

As you can see, the names of these compression fittings key off of what connection the pipe thread is on or the main connection is on (i.e. the branch or the run). Occasionally, you will also run into some female pipe thread on one of these tees.

The reason that you do not mention the composition of the tees all of the time is that with hydraulic fittings, it will be assumed by all those involved that you want brass, bronze, steel, or stainless steel, unless otherwise stated.

Male & Female

The word male or female will be inserted in the description of a fitting if it has male or female pipe thread on one of its connections. Many hydraulic fittings that you will be using have either 3/8 inch [10 DN] or 1/2 inch [15 DN] pipe thread.

Union

In hydraulics the word union is used in the description of a fitting if none of the connections of a fitting has any pipe thread. That is to say, all the connections of the fitting use nuts and ferrules, or all the connections of the fitting are socket weld.

union cross connector
(316 stainless)

The word union in hydraulics means something entirely different than the word union in pipework, so beware. This is one of the instances where a word is spelled the same and pronounced the same, but has two entirely different meanings, depending upon which discipline or trade (hydraulics or pipework) that you happen to be working in at the time.

Flow Control with Integral Check

A very common combination in hydraulics and pneumatics is a variable flow control valve and a check valve hooked in parallel.

7 · Hydraulic Fittings

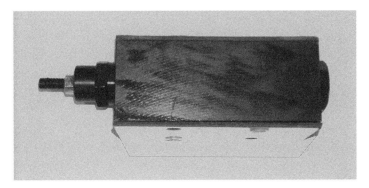

big variable flow control valve

A variable flow control valve and a check valve hooked in parallel inside of one block of material (i.e., one component), is called a variable flow control valve with integral check. This combination allows adjustable (or metered) flow in one direction and free flow in the other.

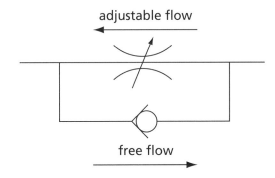

flow control valve with integral check schematic

flow control valves with integral check

You can tell this on the print by looking at the check valve in this combination and determining which way the free flow is through the device. The metered flow (adjustable flow) will then be in the opposite direction. Please note, the leaning arrow does not indicate in which direction you have metered flow.

You can tell most of the time which way the metered flow is by looking on the flow control valve itself for an arrow stamped on the body of the valve. The arrow on the body of the flow control valve points in the direction of adjustable flow. The arrow on the body of a flow control valve with integral check points in the direction of adjustable flow. The arrow on the body of a check valve points in the direction of free flow.

Straight Connector

Most components in hydraulics are connected to hoses or tubing in a circuit using elbows or straight connectors. Straight connectors are specified by stating the pipe thread size and the hose or tubing size.

male straight connectors (37 degree flare)

Pipe Thread

In hydraulics the most common standard pipe thread sizes that you probably will use are 1/2 inch [15 DN] and 3/4 inch [20 DN]. Pipe threads are often confused with the thread on a fitting that the sleeve and nut connect with.

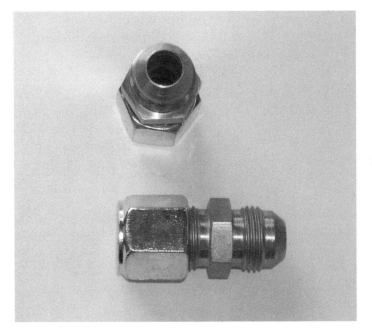

not pipe thread

Sometimes you will pick up fittings from a bin where the nuts and ferrules have fallen off. You can still tell what is or is not pipe thread by looking for the size of the thread. Pipe thread tends to be coarser than the thread for nuts on compression fittings for the same size nominal pipe diameter.

Summary

1) The name of a fitting generally keys off of which connection has pipe thread or the main connection on it.
2) Common pipe thread sizes that you will work with include 1/2 inch [15 DN] and 3/4 inch [20 DN].
3) Common hose sizes that you will work with are 1/2 inch [15 DN] and 5/8 inch [18 DN].
4) Since hose is measured OD and pipe thread is measured ID, initially you will find some confusion in getting your sizes right, visually.
5) Quick disconnect is not always the same as quick connect.

6) Ferrule and sleeve mean the same thing.
7) A union in hydraulics is not the same as a union in pipework.
8) The arrow stamped on the outside of a component means different things on different components.
9) You should order most of your fittings in brass, bronze, steel, stainless steel, or socket weld steel.
10) You should know enough about the standard components called fittings to be able to order them over a telephone or on a computer from a supply house. These items are very common in industry and the average maintenance mechanic should be able to converse with his peers in a knowledgeable manner.

Please turn to page 145 do Exercise #7 before continuing with the text.

Self-Test #2

Do not start the test on the next page until you are ready to begin. Answer all the test questions before looking up any of the answers.

Self-Test #2

(Be Specific)

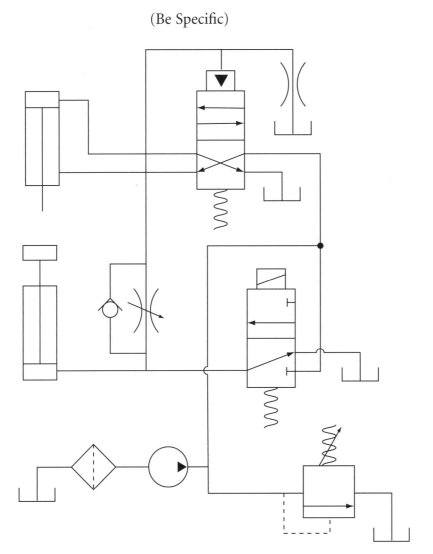

supply is 900 PSIG [6205 kPa]

Note: all items are machined for 3/8" inch [10 DN] pipe thread

1) Which ports would act as return? -10-

2) If the needle valve is adjusted three quarters of the way in, then which actuator moves first? -10-

3) What components would you order to construct three of the circuits in the print above? -70-

4) Solve: force of rod on extension?

ID of actuator = 3 1/2 inch [90 mm]

(line pressure is on print) -10-

Have one of your classmates hand you ten different hydraulic fittings and identify them as if you were ordering them over the phone. No more than one minute for each one.

5) -10-

6) -10-

7) -10-

8) -10-

9) -10-

10) -10-

11) -10-

12) -10-

13) -10-

14) -10-

200 possible points

The Ideal Gas Law and Solved Problems 8

This section of the text is designed to expose the line mechanic to the relationship that exists between pressure, volume, and temperature. This chapter is not necessary to do the job at hand on the floor, but is here for two reasons:

1) to expose the more advanced people in the class (lead line mechanics, etc.) to the temperature, volume, and pressure relationship in order to satisfy their need to know, and
2) to give everyone else a reference to come back to when they need it in the future.

The ideal gas law is given by the following formula:

$$\frac{P_1 V_1}{T_1} = \frac{P_2 V_2}{T_2}$$

where P = absolute pressure (given in PSIA) [kPa]
V = the volume (usually given in cubic feet) [cubic meters]
T = absolute temperature (degrees Rankine) [degrees Kelvin]

The formula says that these three things in situation #1 are equal to these three things in situation #2, hence the subscripts 1 and 2 in the formula. You have already been exposed to the absolute pressure scale, kPa absolute {PSIA}, and volume in cubic meters {cubic feet}, but not the absolute temperature (degrees Rankine or degrees Kelvin) scale. The table on the next page shows how the four systems for describing temperature relate to each other.

Upon successful completion of this unit the student will be able to:
- Record the general concept of the ideal gas law.
- Construct a fluid power circuit using hydraulic actuators.

Learning activities
- Lectures
- Readings
- Demonstrations
- Hands-on experience (Exercise 8)
- Evaluation and
- Discussion of the lab experience.

Temperature Scales

	degrees Fahrenheit (°F)	degrees Rankine (°R)	degrees Centigrade (°C)	degrees Kelvin (°K)
boiling water	212	671	[100]	[373]
freezing water	32	491	[0]	[273]
absolute zero	–459	0	[–273]	[0]

A temperature is really a measurement of how fast a whole bunch of things are hitting you. Since the "things" that we are talking about are so small (atoms and molecules), and so many (billions and trillions and zillions), we perceive temperature as hot and cold. There is an extreme size difference between a human being and a molecule, after all. If a molecule were the size of a dime, a human being would be about the size of the earth. Incomprehensible to say the least. At any rate, there is a point where all molecular motion stops. This point on the temperature scale is called absolute zero and indicates that nothing can get colder (i.e., go slower than be completely stopped).

As you can see from the chart, if you deal in degrees F (Fahrenheit), then the absolute scale you use will be degrees R (Rankine), and if you deal in degrees C [Celsius], then the absolute scale you use will be degrees K [Kelvin].

Let's look at two solved problems using the ideal gas law.

8 · The Ideal Gas Law and Solved Problems

Solved Problem

1) An oxygen tank sitting in a patient's room reads about 2600 PSIG [about 18000 kPa.]. If the bottle contains 8 cubic feet [.2265 cubic meters], how much volume would this represent at STP?

V_1 = 8 cubic feet [.226 cu.m.] V_2 = ?

P_1 = 2615 PSIA [18029 kPa. absolute] P_2 = 14.7 PSIA [101.35 kPa. absolute]

T_1 = 68° F [20°C] T_2 = 68° F [20°C]

$$\frac{P_1 V_1}{T_1} = \frac{P_2 V_2}{T_2}$$

Since $T_1 = T_2$ you do not have to convert to absolute temperature.

then $P_1 V_1 = P_2 V_2$

{substituting standard values}   [substituting metric values]

(2615)(8) = $P_2 V_2$ [18029][.2265] = $P_2 V_2$

(2615)(8) = 14.7 (V_2)   [18029][.2265] = 101.35 (V_2)

20920 = 14.7 (V_2)   [4083] = 101.35 (V_2)

1423 = V_2 [40.3] = V_2

or V_2 = 1423 cubic feet   or V_2 = [40.3 cubic meters]

8 · The Ideal Gas Law and Solved Problems

Solved Problem

2) A 30 cubic foot [.85 cubic meters] receiver reads 100 PSIG [690 kPa.] at 165°F [74°C]. What would it read at 55° F [13°C] several hours later?

Solving using standard values:

V_1 = 30 cubic feet

P_1 = 100 PSIG + 14.7 PSI = 114.7 PSIA

T_1 = 165°F = 624°R

V_2 = 30 cubic feet

P_2 = ?

T_2 = 55°F = 514°R

$$\frac{P_1 V_1}{T_1} = \frac{P_2 V_2}{T_2}$$

since $V_1 = V_2$

$$\frac{P_1}{T_1} = \frac{P_2}{T_2}$$

$$\frac{114.7}{624} = \frac{P_2}{514}$$

Cross multiplying we have:

$(114.7)(514) = 624 (P_2)$

$58956 = 624 (P_2)$

P_2 = 94 PSIA or 79 PSIG

Solving using metric values:

V_1 = .85 cubic meters

P_1 = 690 kPa + 101 kPa = 791 kPa.

T_1 = 74°C = 329°K

V_2 = .85 cubic meters

P_2 = ?

T_2 = 13°C = 268°K

$$\frac{P_1 V_1}{T_1} = \frac{P_2 V_2}{T_2}$$

$$\frac{P_1}{T_1} = \frac{P_2}{T_2}$$

$$\frac{791}{329} = \frac{P_2}{268}$$

$(791)(268) = 329 P_2$

$211988 = 329 P_2$

P_2 = 644 kPa. absolute or 543 kPa. gauge

Please turn to page 148 and do Exercise #8 before continuing with the text.

The Hydraulic Side of Fluid Power

9

Please turn to page 151 and do Exercise #9 before continuing with the text.

Friction

Friction manifests itself as heat in a hydraulic system. In fact, during the old days, the tanks inside plants got so hot that it was not unheard of for someone to set part of their meal on top of the tank so that it was warm and toasty for the morning break or for lunch. When you run a hydraulic unit a lot, you generate "tons" of excess heat. The fluid "thins" out and it has an effect on the operation of your machine. Sometimes you need a cooler to get rid of the extra heat and sometimes you need pressure or temperature compensated flow control valves to overcome the "thinness" of the fluid.

On the other hand, if you have to work outside with a hydraulic unit in cold weather, your fluid is very "thick." This can cause problems with your pump and a lot of your components. You may have to pre-heat the fluid first in order to get everything running okay.

Viscosity

SSU, SAE, SUS, AGMA, [ISO, and centistokes] are all measurements of a fluid's viscosity. For example, at 40 degrees C {104 degrees F} a typical hydraulic fluid at 18 SAE equals approximately 470 SUS, or 2 AGMA, [ISO 68] or [72 centistokes]. The thing to remember is that viscosity for all practical purposes is like the thick-

Upon successful completion of this unit the student will be able to:

- Recognize the advantages of hydraulics over pneumatics in power situations.
- Understand how relevant component choices are determined and how components are sized.
- Construct hydraulic circuits using standard schematic symbols and various types of components.

Learning activities

- Lectures
- Readings
- Demonstrations
- Hands-on experience (Exercise 9)
- Discussion of the lab experience.

ness or the thinness of a fluid, but not always. There are seemingly misleading materials like glycerin. For example, glycerin is a material that feels thin to the touch but has a very high viscosity on the order of honey or hot molasses. Viscosity can easily change 1000% over 60 degrees C {140 degrees F} in real world hydraulic fluids.

Distance, Area, Volume

In hydraulics distance usually shows up as the stroke on a cylinder (i.e., how far the load is moved). Area usually shows up as the surface area of a piston face or a platen on a press. When you are determining how much force that the cylinder or platen exerts, it is equal to F=PA. Force in lbs. [newtons] = Pressure in PSI [Pa.] times Area in sq. inches [sq. meters].

Volume shows up in hydraulics most often as a flow in gallons per minute, GPM [liters per minute, lpm or LPM]. This is basically how much volume of fluid per unit of time that your pump is moving through the hydraulic system. Please remember that on large systems flow can also be given as a mass flow, such as aircraft fuel consumption in kilograms per hour, kg./hr. (you also see this as pounds per hour, lbs./hr.). But, remember the shipping guys. Their shipping documents say that 1 kg. = 2.2 lbs. Not true alert. Not true alert. 1 kg. = 70 slugs. Mass equals mass. Lbs./hr. in this aircraft fuel example is really a fiction (but one that we all pretend to believe in). Mass fuel consumption is really in sph, slugs per hour. In the standard system mass is in slugs (not pounds). The only reason that I keep mentioning this is that in more advanced troubleshooting, it helps to know what the real units are, NOT just what someone on the floor says they are. Do not even begin to talk about slugs to the regular folks on the floor.

Distance, Speed (Velocity), Acceleration

In hydraulics distance usually shows up as the stroke on a cylinder (i.e. how far the load is moved). Speed usually shows up in feet per second, FPS [meters per second, MPS] when the talk turns to how fast your cylinder moves. Of course, when the talk turns to hydraulic motors or pumps, then speed is given in revolutions per minute, RPM. Speed will also show up in advanced troubleshooting when the talk turns to how fast the oil is moving in the hydraulic lines. When speed has a specific direction it is called velocity, but on the floor the two terms are used interchangeably.

Acceleration usually comes into play when the talk turns to how fast the cylinder begins to move from a dead stop up to full speed (advanced troubleshooting). When talking about hydraulic and electric motors, this is usually called ramp up and ramp down.

Flow, Pressure

Flow is generally of three types: volume flow, mass flow, and charge flow. As far as your understanding of troubleshooting is concerned, you can think of these three things as being different manifestations of the same underlying thing, flow. Volume flow is in GPM [lpm or LPM]. Mass flow is in kg./hr. {slugs per hour}. Charge flow, which is also called current, is in amps.

Even though this may seem hard to believe at first, as far as your understanding of troubleshooting is concerned, you can think of pressure and flow as being different manifestations of the same underlying thing, in this book called 4th level tech. terms.

Momentum (a 5th level tech. term)

The concept of momentum is not used very frequently in hydraulics. It is used more in pile drivers, bulletproof vests, and car accidents.

Force, Weight (6th level tech. terms)

Force is most frequently used in hydraulics in the relationship F=PA. The units of force are in pounds [newtons]. A pound is also the unit of weight that we are the most familiar with. Just remember that in the tech. world and in the physics world, 2.2 lbs. = 9.8 newtons, but on the factory floor and outside on the jobsite they use 2.2 lbs. = 1 kg. because of those darned shipping documents.

Energy, Torque (7th level tech. terms)

Joules, BTUs, watts, calories, foot-lbs, and heat are all manifestations of energy.

Power, HP (8th level tech. terms)

The hardest thing for most people to understand is the difference between power and energy. Power is the rate at which energy is used. Say you have a 300-ft. [91-meter] hill behind your town and you run up the hill on Monday. Then on Wednesday you walk up to the top of the same hill. On both days it took exactly the same amount of calories to get to the top of the hill. The same amount of energy. We as human beings perceive power as cardiovascular exercise. Pant, pant.

Distance (L), Period, Inverse Time (a per second or 1/T) are [1st level tech. terms]

Area, Speed (velocity) are [2nd level tech. terms]

Volume (V), Mass (M), Charge (Q) are [3rd level tech. terms]

Flow, Pressure are [4th level tech. terms]

So what this means in intermediate and advanced hydraulic troubleshooting is that, for example, any 4th level tech. term times any 4th level tech. term gives you an 8th level tech. term (i.e., Flow x Pressure = Power), or any 3rd level tech. term times any 3rd level tech. term gives you a 6th level tech. term (i.e.,

Mass x Acceleration = Force), or any 1st level tech. term times any 3rd level tech. term gives you a 4th level tech. term (i.e., Period x Volume = Flow). For most people doing advanced troubleshooting, this way of looking at things is much easier than trying to remember pages and pages of equations from engineering textbooks.

Hydraulic Terms and Concepts That Help Your Understanding

10

Please turn to page 153 and do Exercise #10 before continuing with the text.

Efficiency

Efficiency is always less than 100%. In a system that gives off a lot of heat, your efficiency is low. As time marches on, efficiency is improving. Twenty-five to thirty years ago a car gave off about 85% of its energy as heat, a light bulb had about a 92% loss, and a hydraulic system lost about 45% of its energy as heat. Today a car gives off about 75% of its energy as heat, a light bulb has about a 90% loss, and a hydraulic system loses about 30% as heat. The rule of thumb in hydraulic systems is that for every conversion that you go through, you lose about 10% of your available working circuit energy as heat (it is still at a level of 15% loss per conversion for old systems or poorly designed new systems).

This is how it works today in a hydraulic machine or system:

(1) Electric energy converts to mechanical energy at the electric motor (–10%).

(2) Mechanical energy converts to fluid energy at the pump (–10%).

(3) Fluid energy converts to mechanical energy in the machine (–10%).

Total working circuit energy loss = (30%)
Total heat produced = bunches and bunches

Upon successful completion of this unit the student will be able to:

- Recognize the interrelationships of hydraulic power machines and work situations.
- Understand how relevant circuit choices are determined and how hydraulic units are operated.
- Construct hydraulic circuits using standard schematic symbols and various types of components.

Learning activities

- Lectures
- Readings
- Demonstrations
- Hands-on application (Exercise 10)
- Discussion of the lab experience.

Pressure Differential

If you increase the pressure differential (pressure drop) across a device, it increases the flow. If you decrease the pressure drop across a device, it decreases the flow. When supersonic wind tunnels were first needed to test aircraft and rocket designs, they basically connected two rooms together thorough a small opening that contained the wind tunnel port and pressurized one of the rooms. When the connection between the two rooms slammed open, you got a hell of a lot of air flow in a short period of time through the wind tunnel. The pressure drop was huge.

Dissolved and Entrained Air

Dissolved air in hydraulic fluid is healthy. It is like the air in water that fish "breathe." Two-percent dissolved air in water is what a fish uses. You can't see it, but its there. Nine-percent dissolved air in hydraulic fluid is also healthy. You can't see it, but its there.

Entrained air in hydraulic fluid is unhealthy. It is the air in the hydraulic fluid that is visible, the little bubbles that you can see. They can come from little cracks in the piping system, bad thread fit on your hydraulic components, improper tank design, or too much suction at the pump. If these bubbles get too big they form foam. If these bubbles get too numerous they can cause premature cavitation.

Cavitation

Cavitation occurs at the hydraulic pump when so much vacuum is pulled on the hydraulic fluid at the pump that the oil boils (changes from a liquid to a gas) at room temperature. Remember that high school science lesson where they boiled water two different ways. The first way was by raising the temperature of a beaker of water until the water boiled. The second way was by pulling a vacuum on the beaker of water inside an upside down bell jar. This vacuum was pulled at room temperature until the water started to boil.

How can you tell if a pump is cavitating? Usually by its sound. Say you walk into a room with four pumps running and this is what you hear. Pump #1, hmmm, hmmmmmm, hmmmmmmmm. Pump #2, hmmm, hmmmmmm, hmmmmmmmm. Pump #3, ksdaakk, baackakakakaksdaakk, baackakakakakaksdaakk. Pump #4, hmmm, hmmmmmmm, hmmmmmmmm. Guess which one is cavitating? Pump #3 is.

When a pump is cavitating it is not pumping fluid well or it is not pumping any liquid at all. You will have little bubbles inside the pump that are beating up (exploding against) the inside of the pump housing. It is as if you had a little Smurf inside of the pump housing with a ball peen hammer going to town in a very bad way. On the inside walls of the pump housing you end up having destruction city.

What can you do about it? Well, the number one cause of pump cavitation is a blocked suction line. Start there first. The next two major causes of pump cavitation are improper viscosity of the fluid being moved and incorrectly sized lines in the system for the conditions at hand. Entrained air and a bad pump round out the top five causes.

Actuator Speed, Pressure, Flow, Force, and Torque

The speed of your actuator is a function of flow in GPM [LPM or lpm]. The force of your cylinder is a function of pressure in PSI [kPa.] and area. The torque of your hydraulic motor is a function of pressure in PSI [kPa.] and the speed of the hydraulic fluid through the hydraulic motor.

If you need more oomph, then you go for PRESSURE (the pressure relief valve).

If you want to go faster, then you go for FLOW (the flow control valve).

Unfortunately, in real life there is some overlap when you adjust the pressure relief valve (about 15% or so). As you increase the system pressure, you will also increase the pressure drop across the flow control valves. This will in turn increase the speed of your actuators a bit.

Since this effect is visible, it often leads people working in fluid power to crank up or down on the system pressure to adjust the speed of an actuator, when they should be concentrating almost exclusively on the flow control valves to increase or decrease a cylinder's speed or a hydraulic motor's RPM.

Intensifiers

If you need a large pressure increase on a hydraulic machine for a particular operation {say an 18,000 PSI clamp vs. an 1800 PSI clamp} [124,106 kPa vs. 12,410 kPa], instead of undergoing the huge expense of installing an ultra high pressure multi-stage hydraulic pump, you can install an intensifier. An intensifier looks like a cylinder that has a small additional chamber on the rod end of the cylinder.

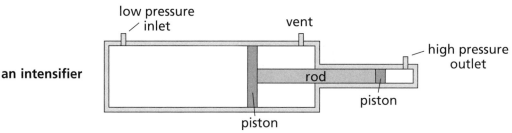

In this example the intensifier has a piston face ratio of 10:1. What that means is if you put oil at 1800 PSI [12,410 kPa] in the inlet port, you will get 18,000 PSI [124,106 kPa] fluid out of the outlet port. This magic can happen if you do not need a high flow of fluid from the outlet port.

Fire Points

If you use oil in a hydraulic unit, you run the risk of possibly starting a fire if a hose bursts, etc. Here is what you need to know about your fluid, whether it is a hydraulic oil or even if it is a fire resistant fluid.

The flash point of a fluid is the temperature at which the fluid is raised to where an open flame (or continuous sparking) will ignite the heated vapors. The fire that results will go out or extinguish if the flame is removed.

The fire point of a fluid is the temperature at which the fluid is raised to where an open flame or continuous sparking will ignite the heated vapors and/or the fluid itself, but the fire that results will NOT go out or extinguish if the flame is removed. This fire can be fought with regular fire extinguishers.

The auto-ignition point of a fluid is the temperature at which the fluid is raised to where it will ignite without an open flame present. The fire that results will NOT go out or extinguish easily. This fire is very hard to fight.

Fire Resistant Fluids

A lot of synthetic fluids are not fire resistant but several are. A lot of water-based fluids do not even have a flash point or a fire point listed because when the fluid gets too hot, the water turns to steam and keeps a fire from starting in the first place. A lot of these same fire resistant fluids still have an auto ignition temperature listed in the specs. For example, the auto ignition point for a water glycol fluid is around 1100 degrees F [593 C], compared to around 600 F [316 C] for oil.

What you need to know is the fire point for the particular fluid that you are using. It is on the MSDS sheet. Know what type of fluid is used in your unit. If you use a water-based fluid, make sure you keep track of any water evaporating out of your fluid so that it can be replaced. And last but not least, if you have a fire, (1) get people out of harm's way, (2) report or call the fire in, and (3) fight the fire if you are allowed to in your work situation.

Hydraulic Cylinder Terminology

Lip seals, bi-direction seals, or piston rings are used to seal around the whole side of the piston in a hydraulic cylinder. This prevents fluid from leaking around the side of a piston when it is moving or when it is still.

piston lip seals

Gland seals are used on the rod end of a hydraulic cylinder to prevent any oil from leaking out of the hydraulic cylinder around the movable rod. They come in several varieties and types, but they are all there for the same reason: prevent any fluid from leaking out around the rod.

rebuild kit for cylinder seals

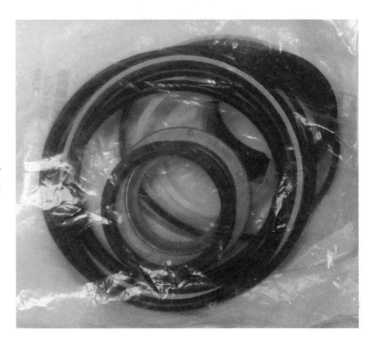

10 · Hydraulic Terms and Concepts

Wiper seals are used on the rod end of a cylinder in order to keep any dirt that may get on the nice shiny cylinder rod from getting inside the cylinder. This wiping action prevents any damage to the rod end gland seals.

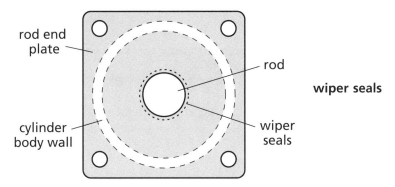

wiper seals

Drains are sometimes used in hydraulic cylinders and are quite frequently used on hydraulic valves to keep any fluid that is trapped between primary and secondary seals from building up pressure and affecting component operations.

Cushions are sometimes built into hydraulic cylinders in order to decrease the likelihood of a shockwave causing problems with blown lines or blown seals. This is done by basically putting a flow control circuit in place within the cylinder that only becomes mechanically activated when the cylinder nears the very end of its stroke. You can adjust the small flow control valve on the cylinder to get just the right end speed. For most of its stroke the cushion has no effect on the speed of the cylinder.

Stop tubes are sometimes used on long cylinders in order to relieve piston rod fulcrum pressure on the rod end gland seals from the lever effect of extended loads. A stop tube resembles a fat cylinder of material that is placed inside a cylinder between the rod end piston face and the internal rod end cylinder wall.

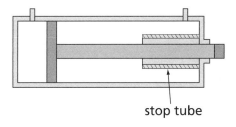

stop tube

A stroke adjuster looks like a giant threaded rod that goes inside of a cylinder with a jam nut on the cap end of a cylinder. This stroke adjuster allows you to have very close control of the actual stroke length of the cylinder in question.

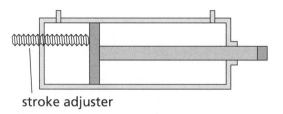

stroke adjuster

Tie rods are very often used on the outside of a hydraulic cylinder in order to physically hold the cap end and the rod end of the cylinder tightly to the cylinder body itself. There are usually four rods.

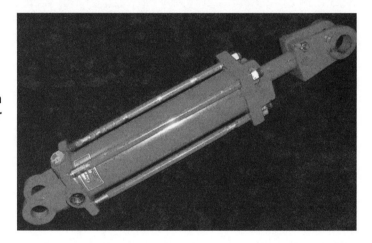

tie rods for a double acting cylinder

The mounting of hydraulic cylinders is accomplished through various mechanical means. Here are some of the most popular.

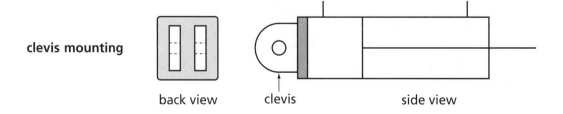

clevis mounting

back view   clevis   side view

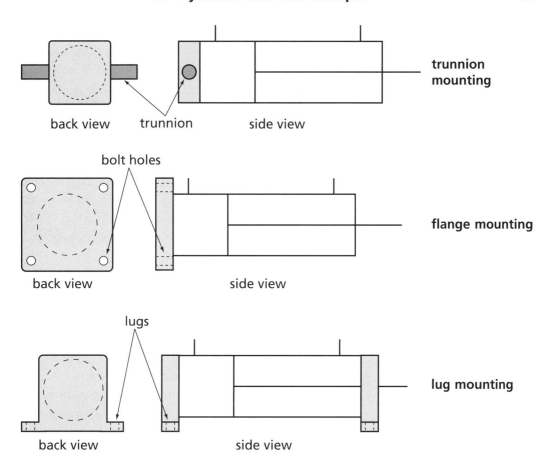

Types of hydraulic cylinders include but are not limited to the following:

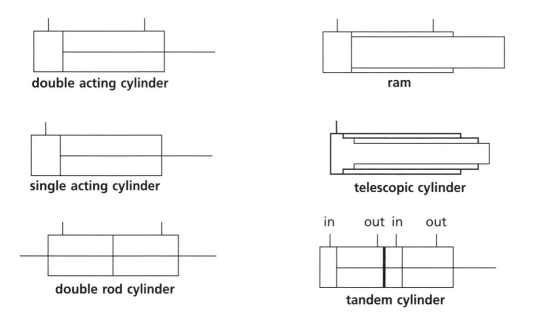

Synchronizing two or more hydraulic cylinders together to do a job is a very difficult thing to do. About the only way to do it successfully in real life is to not try it in the first place. If you really need to do it, use electronically positioned encoded hydraulic cylinders.

Regeneration is a concept sometimes used with hydraulic cylinders where on extension the fluid from the cylinder's rod end is fed back into the cap end of the cylinder along with the normal flow of the pump. This increases the rate of extension of the cylinder without adding another pump to the circuit for more flow. This will be covered in more detail in a later chapter.

Hydraulic Accumulator Terms

Piston type accumulators use a metal piece called a piston to separate the working fluid side from the pressure resisting side (gas, spring, or weight loaded).

Diaphragm type accumulators use a synthetic rubber piece called a diaphragm to separate the working fluid side from the pressure resisting side (nitrogen gas).

Bladder type accumulators use a synthetic rubber piece called a bladder to separate the working fluid side from the pressure resisting side (nitrogen gas is used for oil accumulators, air is used for water accumulators).

Precharge is the name used for the amount of pressure that is left in the pressure resisting side of a gas charged accumulator when all of the fluid is removed from the working fluid side of the accumulator.

Shock absorbing, flow enhancing, and pressure maintaining are the three main reasons that an accumulator is used. For example, to remove a water hammer in a plumbing system or to remove jerking hoses in a hydraulic system, you would use an accumulator to act as a shock absorber.

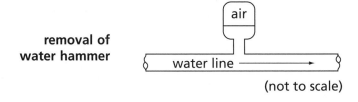

removal of water hammer

(not to scale)

10 · Hydraulic Terms and Concepts

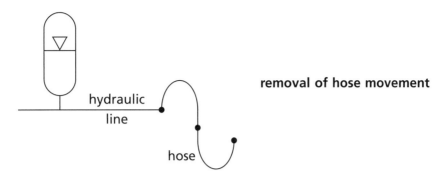

removal of hose movement

To make a cylinder retract faster on a machine, you can use an accumulator to increase flow on retraction. To maintain an even pressure throughout most of your other fluid-based systems, you can use an accumulator called a receiver in pneumatics or a bladder tank in plumbing to mitigate the effects of your compressor (with air) or your well pump (with water).

The ABCs of Hydraulic Relationships

11

Always order a few extra components. You can always use them later, and it will save someone from making an extra trip to the supply house if a small circuit modification is made.

Accumulators are great shock absorbers.

Stacked operator symbols are an **AND** situation.

The nominal and absolute ratings of filters are based on the **Bell shaped curve**.

The number one cause of pump **Cavitation** is a blocked suction line.

In fluid power the **Circle** means something that rotates. The small closed triangle indicates oil flow. The two lines indicate an input and an output. If the triangle points away from the circle or points out of the unit, then it is an oil pump.

Compressibility of a gas (air) is approximately 1700:1, or 1700 to 1, versus the compressibility of a fluid like water or hydraulic oil, which is very nearly 1:1 or 1 to 1.

Dissolved air in hydraulic fluid is healthy.

In actual practice, you will find that **Elbows** are used quite a bit in attaching tubing and hoses to hydraulic components. Most hydraulic lines contain some combination of male elbows and male straight connectors.

Joules, BTUs, watts, calories, foot-lbs, and heat are all manifestations of **Energy**.

Entrained air in hydraulic fluid is unhealthy.

F = PA

> where F = force in pounds [newtons]
> P = pressure in PSI [Pa]
> A = area in square inches [sq. meters]

If you have a **Fire**, (1) get people out of harm's way first, (2) report or call the fire in, and (3) fight the fire if you are allowed to in your work situation.

If you want to adjust the speed of an actuator, then **Go** for the flow control valve.

If you want to adjust the force of a cylinder, then **Go** for the pressure relief valve.

Friction manifests itself as **Heat** in a hydraulic system.

Since hose is measured OD and pipe thread is measured **ID**, initially you will find some confusion in getting your sizes right visually.

746 **Joules** equals 746 newton meters equals 746 watts equals 1 HP.

One **Kilogram** is NOT equal to 2.2 pounds.

Less force (linear motion) = less torque (rotary motion)

Whatever school that you and your co-workers are involved in (meter in or **Meter out**) is okay.

A **Micron** is 10^{-6} meter.

More flow = **More** speed

Remember that in the tech. world and in the Physics world 2.2 lbs. = 9.8 **Newtons**, but on the factory floor and outside on the jobsite they use 2.2 lbs. = 1 kg. because of those shipping documents.

Side by side operator symbols are an **OR** situation.

Power is the rate at which energy is used.

Pressure units are [Pascals], [kPa], [MPa], [bar], or pounds per square inch {PSI}.

inside view of a pressure gauge

PTAB stands for pressure port (P), tank port (T), actuator port (A), and actuator port (B). If you reverse the A and B ports your actuator runs backwards.

Quiet and steady troubleshooting will win the day.

Pressure = flow x **Resistance**

You need to mention two points and the units to define a **Scale**.

Remember, all **Tanks** are reservoirs, but not all reservoirs are tanks.

If you have an accident and get caught in a hydraulic system (liquid), most of the time you will NOT get your fingers, etc. out until the machine cycles again because the cylinder has no give (no compressibility factor). A hydraulic system is very **Unforgiving** to the human anatomy.

The leaning arrow stands for the word **Variable** when it is superimposed on top of another schematic symbol.

Viscosity for all practical purposes is the thickness or the thinness of a fluid.

(**Watts**) = (amps) x (volts)

A **Weighted** platen can lower a load using a single acting cylinder and gravity.

Power = Flow **X** Pressure

You generally work in either standard or metric units.

Zero tolerance for horseplay.

Hydraulic Motors 12

Hydraulic motor terminology can be a bit confusing, so I am going to give you the company line first, and then tell you an easier way to remember it.

Positive Displacement Hydraulic Motors

Most of the hydraulic motors that you run into are of the positive displacement type. What this means is that the amount of fluid going through the motor chamber is a constant (fixed displacement) because the internal chamber of the motor remains the same size. Generally, the only variables are the speed of the motor (controlled by a flow control valve) and the torque of the motor at constant flow (controlled by a pressure relief valve). The main types of positive displacement motors are:

Vane Motors

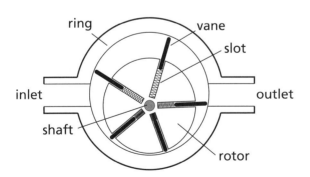

cutaway view of a vane motor

Upon successful completion of this unit the student will be able to:

- Discuss types of hydraulic motors and how they operate.
- Gain additional insight into torque, pressure, flow, area, and force of hydraulic pumps and hydraulic motors.
- Construct additional types of hydraulic circuits.

Learning activities

- Lectures
- Readings
- Demonstrations
- Exercise #11

12 · Hydraulic Motors

Gear Motors

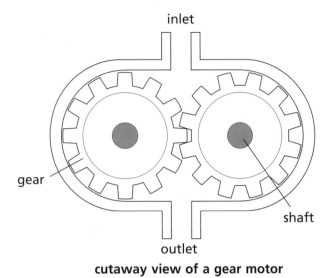

cutaway view of a gear motor

Piston Motors

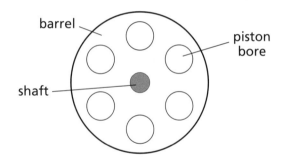

cutaway view of a piston motor

Variable Displacement Hydraulic Motors

Some of the hydraulic motors that you run into are of the variable displacement type. What this means is that the amount of fluid going through the motor chamber is a constant if the internal chamber of the motor remains the same size. That would be true until it is adjusted (variable displacement). The three variables are generally (1) the speed and the torque of the motor (the adjustment of the swashplate angle of the motor that affects both), (2) the speed of the motor (controlled by a flow control valve), and (3) the torque of the motor at constant flow (controlled by a pressure relief valve). The main type of variable displacement motor is the:

Axial Piston Swashplate Motor

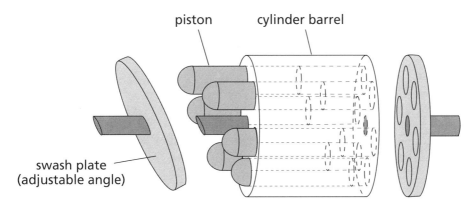

exploded view of swashplate and pistons

The easy way to remember all of this is that if the hydraulic motor is adjustable, it is called a variable displacement hydraulic motor, and if the hydraulic motor itself is not adjustable, it is called a positive displacement hydraulic motor.

RPM

The RPM of a hydraulic motor is equal to the hydraulic flow divided by the motor displacement. It is relatively easy to get the RPM information from a tachometer, the GPM information from a flow meter, and the motor displacement information from a spec. sheet. You would have to calculate the motor displacement if you did not have the original spec. sheet for the motor. The formula method is listed below for informational purposes.

motor RPM = flow/motor displacement
or
motor displacement = flow/motor RPM

12 · Hydraulic Motors

for example:

motor RPM = 500 RPM
flow = 10 GPM [37.85 liters/min. or LPM]

motor displacement = flow/RPM

= 10 GPM/500 RPM   [37.85 LPM/500 RPM]

= 10(231 cu. inches/min.)   [(37850 cc/min.)
      (500/min.)   (500/min.)]

= 4.6 cu. inches   [76 cc]

Torque

In order to get more speed out of your hydraulic motor, crank up the flow control valve. In order to get more torque out of your hydraulic motor at constant flow, crank up the pressure relief valve.

The torque of a hydraulic motor is equal to the hydraulic motor horsepower divided by the motor RPM. Get the RPM information from a tachometer, the horsepower information from a spec. sheet or from the physical size of the hydraulic motor, and the torque information from a spec. sheet or from a formula. You would have to calculate the motor torque if you did not have the original spec. sheet for the motor.

motor torque = motor HP/motor RPM
or
motor torque = HP/RPM

for example: motor RPM = 500 RPM
motor HP = 5 HP [3.73 kW]

motor torque = HP/RPM [kW/RPM]

= 5 HP/500 RPM   [3.73 kW/500 RPM]

= 5(550 ft.lbs/sec.)   [(3730 watts/sec.)
  (500/min.)   (500/min.)]

= 5(33,000 ft.lbs/min.)   [(223,800 watts/min.)
  (500/min.)   (500/min)]

                              [(223,800 joules/min)
                                (500/min)]

= 165,000 ft.lbs/min.   [(223,800 Nt.m/min.)
     (500/min.)   (500/min.)]

= 330 ft.lbs   [447 Nt.m]

[for the maintenance mechanic and shipping guys that's 45.5 kg.m on the torque wrench pseudo scale]

Horsepower

The most important thing to remember when dealing with the horsepower rating of a hydraulic motor is the almost unbelievable weight to horsepower ratio of a hydraulic motor compared to an electric one. It's an astounding 20 to 1. In other words, on average, a 10 HP electric motor weighs 20 times more than a 10 HP hydraulic motor.

Imagine the possibilities in trucks or automobiles for starters.

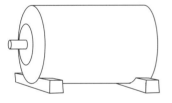

Weight to Horsepower Ratio 20:1

10 hp electric motor   10 hp hydraulic motor

Freewheeling

When a load attached to a hydraulic motor is allowed to coast, it is said to be freewheeling.

Please use the internet to look up the following items and terms. The two best sources that I use as of right now are the article library available at the Fluid Power Journal (the official publication of the Fluid Power Society) at www.fluidpowerjournal.com and the Google™ search engine.

12 · Hydraulic Motors

Items & Terms:

Hydraulic Motors

Hydraulic Pumps

Power = Pressure times Flow

Flow gives you Speed or RPM

Pressure gives you Force

Pressure at constant flow gives you Torque

Turn to page 156 and do Exercise #11. You will be building a circuit to demonstrate at least two of the above items or terms. Explain your results to your instructor and to your class.

Do not start the test on the next page until you are ready to begin. Answer all the test questions before looking up any of the answers.

Self-Test #3

As regards the print below:

1) What happens if #V8 is activated? -25-

2) Describe #V13. -15-

3) Is #f2 metering in or metering out? Why or why not? -15-

4) If #V13 is activated and #V3 is energized, then what happens? -15-

5) Which actuator retracts at full speed? Why? -15-

Be specific.
85 possible points

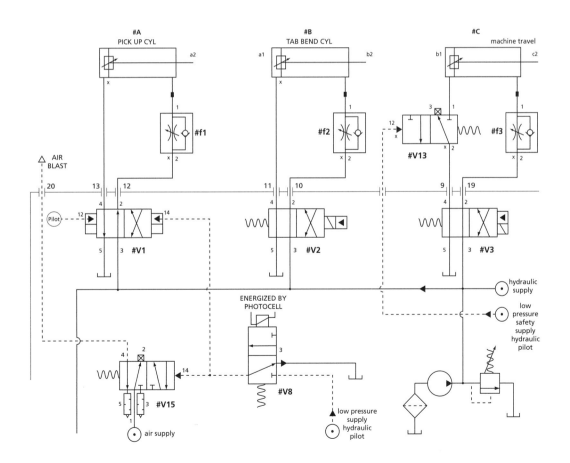

Pressure Control Valves 13

Please turn to page 159 and do Exercise #12 before continuing with the text.

One of the main aspects of hydraulics that is different from pneumatics is the compressibility of the fluid. Since air is highly compressible, most pressure actuated control valves in pneumatics are subject to false triggering. This limits the attempted use of pressure as a means of control in the pneumatic fluid power area.

That is not the case with hydraulics. Since oil is very incompressible, most pressure actuated control valves in hydraulics work fine. This opens up the use of pressure as a means of control in the hydraulic fluid power area. This is true not only in the area of pressure control valves, but also in the area of cylinder rod positioning using encoders.

There is an entire family of pressure control valves in hydraulics. They are normally closed,

normally closed pressure control valve

or they are normally open.

normally open pressure control valve

The distinguishing characteristics of these valves are that (1) they change state from a mechanically adjustable "pressure trigger point also called a set point"

Upon successful completion of this unit the student will be able to:

- Recognize the makeup of the units and the sequence of operation of the major types of pressure control valves.
- Understand how to size components.
- Construct hydraulic circuits using standard schematic symbols and various types of pressure control valves.

Learning activities

- Lectures
- Readings
- Demonstrations
- Hands-on applications (Exercises #12–15)
- Discussion of the lab experience

and that (2) they use remote pilots as well as internal pilots. The most common pressure control valves have a family tree that looks something like this:

N.C. (normally closed) pressure control valves

pressure relief valve
sequence valve
brake valve
counterbalance valve
unloading valve

N.O. (normally open) pressure control valves

pressure reducing valve

Let us take a look at these major categories of valves one by one.

Pressure Relief Valve

The pressure relief valve is an N.C. pressure control valve that is used in most instances to set your hydraulic unit's pressure. In the following example, the set point is 900 PSIG [6205 kPa]. The system pressure maximum that could be reached at the pump outlet if you deadheaded the pump output (closed off the rest of the circuit completely at the pump outlet) is 1100 PSIG [7584 kPa].

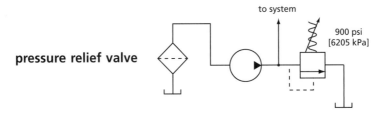

Sequence Valve

The sequence valve is an N.C. pressure control valve that is used to do a one-two punch (i.e., move an actuator after another actuator has already moved). In the following example, the set point of the sequence valve is 500 PSIG [3447 kPa] and the system pressure is 900 PSIG [6205 kPa]. In this circuit cylinder #1 moves first, then cylinder #2 moves next. Do you know why?

13 · Pressure Control Valves

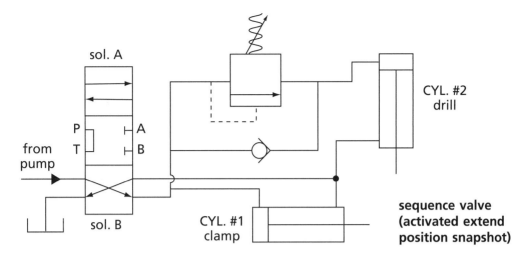

sequence valve (activated extend position snapshot)

Please turn to page 163 and do Exercise #13 (the clamp and drill sequence valve circuit) before continuing with the text.

Brake Valve

The brake valve is an N.C. pressure control valve that is used to have one mechanical set point tripped via an internal pilot for normal hydraulic motor operation, and another set point tripped via an external pilot to control any runaway loads. In the following example, the set point of the brake valve is 700 PSIG [4826 kPa], the system pressure is 900 PSIG [6205 kPa], and the remote pilot operates around 100 PSIG [689 kPa]. In this particular circuit the load rotates only clockwise (CW).

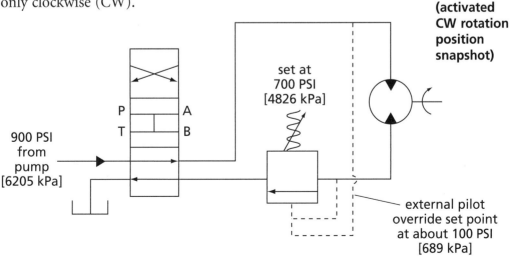

brake valve (activated CW rotation position snapshot)

Counterbalance Valve

The counterbalance valve is an N.C. pressure control valve that is used for normal hydraulic platen operation so that the load does not drop down too fast and crash into something. The valve may also be used to control any runaway loads from a hydraulic motor. In the following example, the set point of the counterbalance valve is 700 PSIG [4826 kPa] and the system pressure is 900 PSIG [6205 kPa]. In this particular circuit the platen retracts up and down at full speed, but the platen is not allowed to "drop" too fast on extension due to the placement of the counterbalance valve in the circuit.

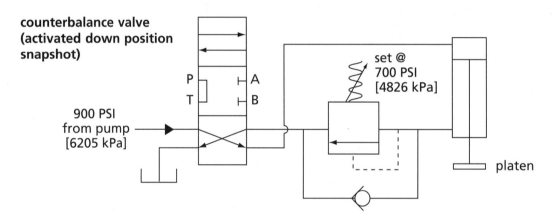

Unloading Valve

The unloading valve is an N.C. pressure control valve that is used for normal hydraulic actuator operation so that the pressure at an actuator in the circuit has a maximum pressure cutoff at some predetermined setting. The unloading valve is usually used in conjunction with a check valve and with an external pilot line. In the following example, the set point of the unloading valve is 700 PSIG [4826 kPa] and the system pressure maximum that could be reached at the pump outlet is 1100 PSIG [7584 kPa]. In this particular circuit a cylinder would be able to retract and extend at full speed at no more than a maximum of 700 PSIG [4826 kPa] of pressure.

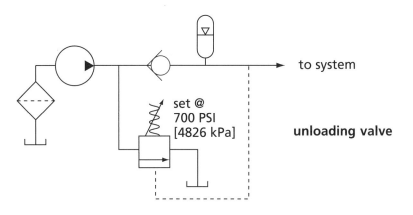

unloading valve

Pressure Reducing Valve

The pressure reducing valve is an N.O. pressure control valve that is used for normal hydraulic actuator operation so that the pressure at one or more of the actuators in the circuit is different than the pressure at the other actuators. The pressure reducer valve is usually used in conjunction with clamping circuits, or in some instances with selective torque reduction of hydraulic motors. In the following example, the set point of the pressure reducer valve is 500 PSIG [3447 kPa] and the pressure maximum that could be reached at the pump outlet is 1100 PSIG [7584 kPa], while the system pressure is set at 900 PSIG [6205 kPa]. In this particular circuit cylinders A and B extend. Cylinder A extends and then clamps at 900 PSI [6205 kPa]. Cylinder B extends at the same time but clamps at 500 PSI [3447 kPa].

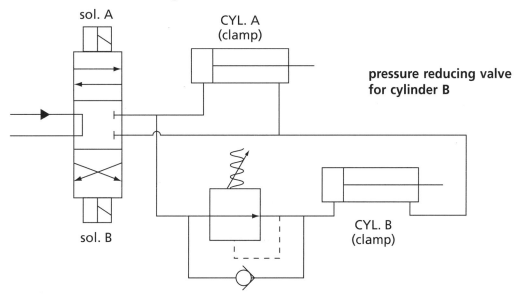

pressure reducing valve for cylinder B

13 · Pressure Control Valves

Please turn to page 166 and do Exercise #14 (the pressure reducing valve circuit) before continuing with the text.

The use of pressure control valves in hydraulic circuitry gives you whole other level of control over fluid power circuits that pneumatic circuits lack when dealing with many aspects of fluid power loads.

Summary

To summarize, the main facts of pressure control valves are:

1) They are N.O. and N.C.
2) Most of them are N.C.
3) Their name keys off of how they are used in a circuit.
4) Most use mechanical means to set pressure trip points.
5) An external pilot that is activated on a pressure control valve will override the mechanical set point. (This in effect gives you two or even three possible pressure trip points on one valve.)
6) These valves can have different combinations of internal and external pilot lines to give you the type of operation that you need.
7) And last, but not least, in an emergency situation you can use a pressure relief valve with an external pilot port to substitute for just about any other type of pressure control valve.

Please turn to page 169 and do Exercise #15, (the pressure relief valve revisited circuit). This particular circuit will show you how to get three pressure trip points out of one valve. The important thing to remember about this is that any hydraulic pressure control valve that has an external pilot line can have two or three different trip points (actuation points). **Any external pilot line pressure to a pressure relief valve lower in value than the mechanically set trip point of that pressure relief valve and above the value of the unloaded pressure relief valve (typically 100–150 PSI [689–1034 kPa] or so) will override the set point and cause that pressure relief valve to actuate.**

Proportional and Servo Valves 14

Please turn to page 166 and do Exercise #16 before continuing with the text.

One of the main technical aspects of hydraulics in the fluid power world is that higher levels of position control are possible due to fluid compressibility. Since oil is very incompressible, most pressure actuated control valves in hydraulics work fine, but the next level of position control in hydraulics is with proportional and servo valves. If you can (1) keep your fluid super clean, (2) calculate set points (e.g., proportional valves) or hook up appropriate encoders (e.g., servos) for your particular work environment, and (3) have a general understanding of open loop and closed loop control, then proportional valves and servos are for you. This opens up the use of pressure and flow as a means of control in the hydraulic fluid power area to a much greater degree than pressure control valves can obtain by themselves.

Six General Areas of Valve Control in Hydraulics

Electro-mechanical control. Typically in the midpoint areas of 1/10 GPM to 10 GPM [.38 LPM to 38 LPM] flow rates and 5 Hz. to 500 Hz. valve actuation rates.

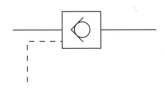

**electro-mechanical actuation control
(pilot operated check valve)**

Upon successful completion of this unit the student will be able to:

- Recognize the makeup of the units and the sequence of operation of the major types of proportional and servo valves.
- Understand how to size components.
- Construct hydraulic circuits using standard schematic symbols and various types of valve symbols.

Learning activities

- Lectures
- Readings
- Demonstrations
- Hands-on hydraulic circuit applications (Exercise #16)
- Discussion of the lab experience

Hydrostatic control, also called hydrostatic drive control. Typically in the midpoint areas of 20 GPM to 200 GPM [76 LPM to 760 LPM] flow rates and 5 Hz. to 25 Hz. valve actuation rates.

bottom view of a lawn mower's hydrostatic drive unit

Low to medium power, low to mid frequency control, also called proportional valve control. Typically in the midpoint areas of 1/10 GPM to 50 GPM [.38 LPM to 190 LPM] flow rates and 5 Hz. to 200 Hz. valve actuation rates. Often found in cartridge valve format.

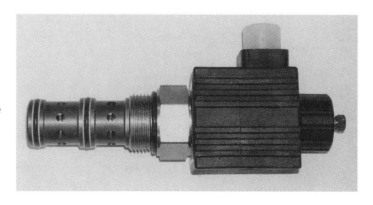

cartridge valve

14 · Proportional and Servo Valves

Low power, high frequency control, also called single stage and/or instrument servo control. Typically in the midpoint areas of 1/10 GPM to 1 GPM [.38 LPM to 3.8 LPM] flow rates and 500 Hz. to 1200 Hz. valve actuation rates.

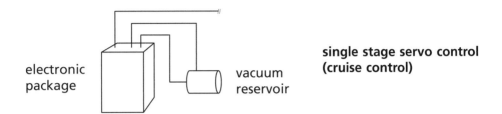

single stage servo control (cruise control)

High power, broad spectrum frequency control also called two stage servo control. Typically in the midpoint areas of .4 GPM to 50 GPM [1.5 LPM to 190 LPM] flow rates and 5 Hz. to 1200 Hz. valve actuation rates.

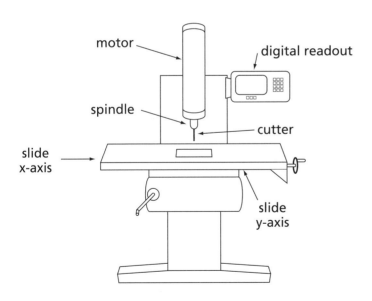

two stage servo control (CNC mill)

14 · Proportional and Servo Valves

High power, lower frequency control, also called three stage servo control. Typically in the midpoint areas of 40 GPM to 500 GPM [151 LPM to 1890 LPM] flow rates and 5 Hz. to 200 Hz. valve actuation rates.

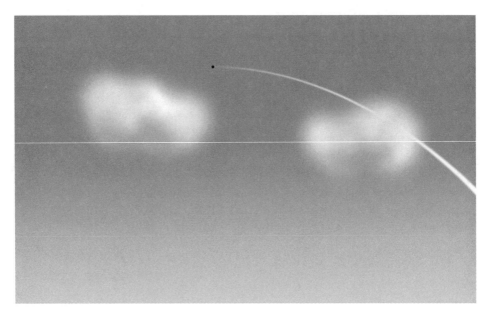

Look, up in the sky, it's a bird, it's a plane, it's three stage servo control.

The distinguishing characteristics of servo valves are (1) they trigger off of an electrical or electronic feedback loop, also called a "feedback circuit," and (2) they use external sensors to set a tripping point that causes the valve spool or valve mechanism to actuate back and forth to maintain control of the actuators in the circuit. The most common servo valves in hydraulics have a family tree that looks something like this:

single stage and/or instrument servo control
laser positioning
rocket fin positioning
radar magnetron tuning
instrumentation positioning
wire EDM

two stage servo control
machine tools
robots
process control
turbine control
military gun positioning
missile systems
blow molding
anti skid braking

three stage servo control
flight simulators
primary flight controls
tank systems
injection molding
roller mills
die casting
hydrofoils

Let us take a closer look at these families one by one by looking at somewhat familiar examples of each.

Electro-mechanical Control (Levers and Solenoids)

Lever and solenoid operated valves can be thought of as your typical on-off type circuit situation most of the time. As in, I punch the button and the cardboard boxes get crushed in the bailing machine. However, some people are very adept at feathering using a lever. An example of feathering would be lowering or raising a load on a forklift very gently while positioning it carefully in place in a storage rack.

Hydrostatic Drive Control (Hydrostatic Drives)

Hydrostatic drives are going to be in a lot of construction equipment and high-end lawn tractors that you will be around. The hydrostatic drive is basically a hydraulic pump hooked to a hydraulic motor within the same housing or not. The distinguishing characteristic of the hydrostatic drive is that you can have a continuous (non-jerky) application of power and torque from your engine to your wheels. It takes shifting gears out of the equation almost all of the time.

hydrostatic drive

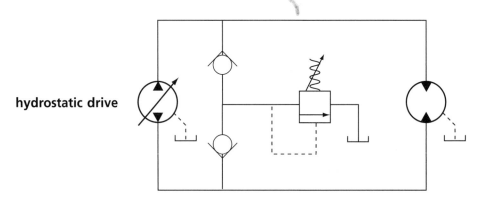

Proportional Valve Control

The distinguishing characteristic of a proportional valve is that it can open or shut a valve stem part of the way instead of just full ON or full OFF. That, of course, will give you an infinite variety of flow settings for the application in your circuit. There are two main types of proportional control valves that you run into in maintenance work.

The first types are proportional valves that are controlling fluid flows that are relatively big. These are the ones that you see controlling a large water glycol line in refrigeration work, air flow in HVAC and production equipment, water flow in hydronic systems, gas line flow into a giant oven, etc. They are controlled by a variety of means, but all are called proportional valves or proportional something or other.

The second type is the one that you find in hydraulics. These proportional valves resemble a regular directional control valve on a print, but they allow you to infinitely position the valve spool so that you can have different levels of flow going through the directional control valve port. These valves use DC coils and use a varying current at a constant voltage to position the valve stem.

14 · Proportional and Servo Valves

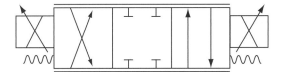

solenoid operated, infinite position,
4 way, spring centered proportional
directional control valve

Single Stage and/or Instrument Servo Control

This type of servo control is typically found in lighter duty control applications. Your computer-based printers or scanners, for example. This is where the print head in the case of printers or the photo detector in the case of scanners must be positioned by stepper motor(s) of some sort into a very rigorously defined position. Stepper motors are controlled by feedback sensors and software programming.

electric servo motors for a printer (top cogged belt is for horizontal positioning servo, bottom left servo is for vertical positioning)

In hydraulics, single stage servos are used in control applications where proportional valves were used in the past because the single stage servo is cheaper and is more easily controlled by off the shelf electronic components. Directional control valves are controlled by a servo with feedback sensors and software programming.

single stage servo valve (outside view)

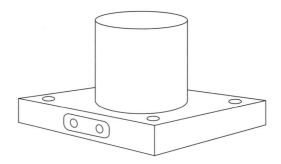

Two Stage Servo Control

This type of servo control is typically found in heavier duty control applications where much more energy, torque, and force are involved. Your car is a good example. We are not talking about the wimpy cruise control servo, but rather the robust anti-lock brake servo. The brakes are controlled by a two stage servo in order to position your brake shoes and brake calipers into very rigorously defined positions, very quickly if needed. Brakes are quickly controlled by feedback sensors and software programming if stuff starts going to hell in a hand basket on slick pavement or on an icy road surface. In this hydraulic system, two stage servos are used in a control application to increase safety significantly, where only foot control and a driver's knowledge, or lack thereof, was used in the past.

Heavier energy situations are what require two stage servos. The car's anti-lock brake servo control is used to rapidly pump the brakes (a heavy energy component) when the need arises in order to keep you from skidding. In a packaging situation, the two stage servo control is used to accurately position a product quickly to a particular station in a particular orientation (a high energy situation) to move, cap, label, code date, load, etc. In commercial printing, two stage servos are used to very accurately track registration of images during printing at high speeds (another higher energy situation). It's almost unbelievable how hard it is to stop or start a full roll of paper quickly.

14 · Proportional and Servo Valves

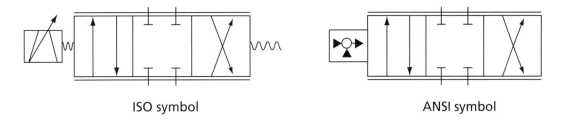

ISO symbol   ANSI symbol

Three Stage Servo Control

This type of servo control is typically found in very heavy duty control applications where much, much energy, torque, force, response time, and speed or combinations thereof are involved. Your vacation trip is a good example. The primary flight control systems of the jet you are on are controlled by three stage servos in order to position the control surfaces of the plane into very rigorously defined positions very quickly if needed or very slowly if everything is normal. Control surfaces of the plane are quickly controlled by feedback sensors and software programming if stuff starts getting too fast for the pilot to react, but at the same time these same three stage servos are used for a very light touch when the autopilot is on, in straight and level flight. In this airborne hydraulic system, three stage servos are used in a control application to increase safety and comfort, and to get good "gas" mileage.

In a flight simulator situation where pilots are trained on good old Mother Earth, three stage servos are also used. They are used to accurately position the mock cockpit. This is done either slowly or quickly under computer control in order to trick a person in a particular situation (flight scenario) into thinking that they are in a particular physical orientation (such as getting ready to crash into the old fort on approach to PHL, falling down out of the sky in a microburst, etc.). In flight simulation, three stage servo motors are used to very accurately fool the pilot's senses during simulated maneuvers at high or low speeds. After a few minutes it makes you wonder if your feet are still on the ground.

Discussion of Self-Test Answers 15

Test #1

1) a. oil filter (wet), often called a hydraulic breather tube strainer or filter, usually located in the throat of the filler tube
   b. oil filter (dry), often called a hydraulic breather tube air filter, usually located in the cap of the filler tube
   c. adsorption (if needed)
   d. suction line filter
   e. regulator
   f. hydraulic oil cooler
   g. separation
   h. pressure line filter or return line filter
   i. in tank strainer

Separation can come before or after the hydraulic cooler or both. Either is acceptable. You need to mention **ad**sorption in (b), not **ab**sorbtion.

2) hyd.—ring or loop (see text about pseudo branch)
   pneu.—ring or loop, multi-branch, single outlet

3) solid   plasma
   liquid   gas
   gas   liquid
   plasma   solid

You can list them either way, as long as they are in the proper order.

4) In your definition you need to mention at least these two things:
   a. gives up or takes on heat and
   b. the temperature remains the same

5) In your definition you need to mention that the ratio of the compressibility of a fluid (like water or hydraulic oil) is 1:1 or 1 to 1, while the ratio of the compressibility of a gas (air) is approximately 1700:1, or 1700 to 1.

6) The point that free air is the air all around you (a human being) wherever you happen to be needs to be made in your definition. This is the air in the headspace of the hydraulic tank.

7) 14.7 PSIA [101.3 kPa absolute], 68 degrees F [20 degrees C], 36% R.H.

8) This is generally the hardest definition to put in words, and many times a student will struggle with it. The points that need to be touched on are:

   a. the R.H. measurement is a percent

   b. the percent is based on the maximum amount of water vapor that the headspace in the oil tank can hold at a particular temperature (complete saturation of the air in the headspace would be 100% R.H.)

   c. the actual amount of water that a particular volume of headspace air can hold varies with temperature

9) "Mechanical" is referring to the hydraulic system piping and actuator rods in this instance. Your answer should have pitch, tubing, hoses, drains, boots, wipers, etc. in it (see text illustrations).

10) You need to mention two points and the units to define a scale.

    units = PSI (pounds per square inch) [kPa]

    one point on the scale = 0 PSIA [0 kPa absolute] a perfect vacuum

    another point on the scale = 14.7 PSIA [101.3 kPa absolute] sea level pressure

11) a. 3 position, 4 way, tandem center, solenoid operated, spring centered, directional control valve

    b. 2 position, 3 way, hydraulic pilot operated OR manually operated, spring return, directional control valve

15 · Discussion of Self Test Answers

   c. single acting cylinder
   d. fixed orifice flow control valve
   e. This stands for the word variable when it is superimposed on top of another schematic symbol.
   f. a check valve
   g. tank

   Remember, all tanks are reservoirs, but not all reservoirs are tanks.

   h. oil treatment (specifically an oil filter)

   At this point in time, all that you need to recognize is that the general diamond shape on the print stands for some type of oil treatment.

   i. variable flow control valve
   j. This answer comes from a little detective work. The circle means something that rotates. The small closed triangle indicates oil flow. The two lines indicate an input and an output (or intake and output). The triangle points away from the circle or indicates flow out of the unit. You guessed it! An oil pump.
   k. a gas charged accumulator

   Remember, all accumulators are reservoirs, but not all reservoirs are accumulators.

   l. double acting cylinder

   See, it has two ports versus the one port the single acting cylinder has in (c).

   m. lines connected
   n. two lines crossing, not connected
   o. hydraulic flow or hydraulic line
   p. variable flow control valve with integral check
   q. enclosure line

   You can tell this because the dotted line closes back in on itself.

12) In your definition you need to mention that:
   a. the dew point is the particular temperature that a given volume of headspace air above the oil

in the tank reaches where the relative humidity will be 100%, and

b. at the dew point you have a change of state, water vapor to liquid water.

13) units = PSI [kPa]

one point on the scale is 0 PSIG [0 kPa gauge] = atmospheric pressure

another point on the scale is 900 to 1100 PSIG [6200–7600 kPa gauge] = a typical hydraulic system pressure

14) oil, water glycol, oil in water, water in oil, or artificial

It usually depends on the fire resistance needed and the temperature that the machine operates in.

15) units = "Hg (inches of mercury)

one point on the scale is 29.92" Hg [101.3 kPa vacuum scale] = a perfect vacuum

another point on the scale is 0" Hg [0 kPa vacuum scale] = sea level pressure

16) ⟵ Look at the check valve, not at the leaning arrow.

17) There is no free flow. You have metered flow in either direction.

18) Standard Temperature and Pressure

19) reversible hydraulic motor

This one is hard the first time through. It is one that you need to memorize.

15 · Discussion of Self Test Answers

Test #2

1)

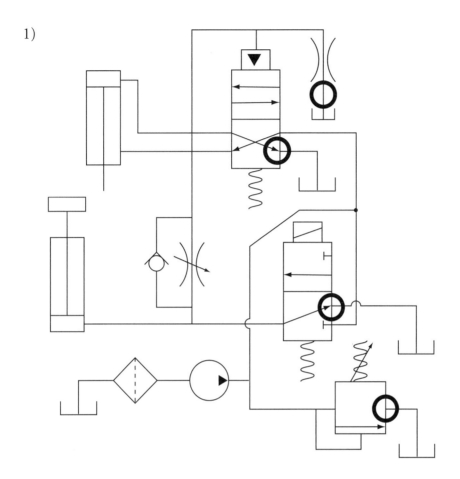

2) There are two acceptable answers to this question, depending on how you look at it.

   a. normal operating conditions (the solenoid operated directional control valve is triggered) The upright double acting cylinder starts to move first and after some delay (determined by the setting of the variable flow control valve), the downward facing double acting cylinder moves.

   b. the solenoid operated directional control valve is not triggered Neither cylinder moves first because none of them will move at all.

3) three, 2 position, 3 way, solenoid operated, spring return, hydraulic directional control valves

three, 2 position, 4 way, hydraulic pilot operated, spring return, directional control valves

three variable flow control valves with integral check

three fixed orifice flow control valves

three pressure relief valves

three suction line filters

three hydraulic pumps

three hydraulic tanks

three single acting cylinders, 2 inch [50 DN] bore, 12 inch [300 mm] stroke (the actual bore and stroke may vary on your particular order of course, but you should include these two items at this time)

three double acting cylinders, 2 inch [50 DN] bore, 12 inch [300 mm] stroke (ditto)

100 ft. roll of 1/2" [15 DN] hydraulic hose

sixty connectors, 1/2" [15 DN] tubing, 3/8" [10 DN] pipe thread

In actual practice, you will find that elbows are used quite a bit in attaching tubing to hydraulic components, so that this particular line item would contain some combination of male elbows and straight connectors. The actual number of fittings needed for this circuit is around fifty four.

eight male branch tees, 1/2" [15 DN] tubing,

3/8" [10 DN] pipe thread

Always order a few extra. You can always use them later, and it will save someone from making an extra trip to the supply house if a small circuit modification is made.

15 · Discussion of Self Test Answers

4) given ID = 3 1/2 inches [90 DN]

pressure = 900 PSIG [6205 kPa] (on print)

solve for force

$$F = PA$$

where F = force in pounds   [in newtons]

P = pressure in PSI   [Pascals]

A = area in square inches   [square meters]

substituting we have:

F = (900)A   eqn. 4.1   [F = 6,205,281 Pa(A)]

solving for A we have:

$A = \pi r^2$ eqn. 4.2

the ID = 3 1/2 inches [89 mm] so the radius will equal one half of this or 1.75 inches [44.45 mm]

Substituting in equation 4.2 we have:

A = (π) (1.75)(1.75)   [A = (pi) (44.45)(44.45)]

A = (3.1416)(1.75)(1.75)   [A = (3.1416) (44.45)(44.45)]

A = (3.1416)(3.063)   [A = (3.1416) (1976)]

A = 9.623 square inches   [A = 6208 square mm or

.006208 square meters]

substituting in equation 4.1 we have:

F = (900)(9.623)   [F = (6205281)(.006208)]

F = 8661 lbs.   [F = 38,526 newtons]

The force on the rod on extension will be about 8661 pounds [38,526 newtons].

15 · Discussion of Self Test Answers

Test #3

Refer to the print on the following page:

1) If #V8 is activated, then:

   a. directional control valve #V15 is activated and the air blast (upper left hand corner) comes on,

   b. directional control valve #V1 is activated and cylinder #A extends.

2) #V13 is a two position, three way, hydraulic pilot operated, spring return, directional control valve that has one port plugged.

3) #F2 is metering out. The term meter in or meter out in hydraulics refers to using a flow control valve to meter the flow going into (meter in) or coming out (meter out) of an actuator. Since on cylinder #B the flow control valve just meters the flow going out of the cylinder on extension, it is a meter out situation. Again, in most factories they don't care which school you are in (meter in or meter out). They just want you to be able to tell the difference between the two so that you can troubleshoot the circuit.

4) Cylinder #C extends. Remember, the print shows all directional control valves in their unenergized, deactivated, or rest position.

5) #A, #B, and #C. All three cylinders retract at full speed. All three cylinders will have free flow through their respective flow control valves on retraction.

15 · Discussion of Self Test Answers

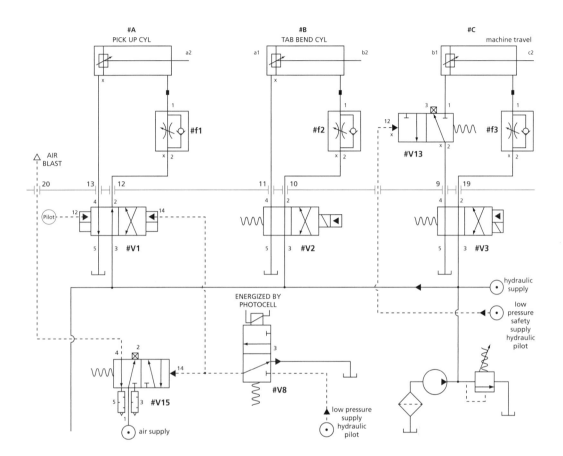

Exercises 16

All of the textbook exercises are listed in order (#1 through #16) at the end of this chapter. The order that you should do the reviews and the exercises for this chapter is listed out below as Roman numerals I through IV; do this chapter in this order if you have come here first and skipped any of the text:

I. Review Chapter 11 on the "The ABCs of Hydraulic Relationships."

II. Review all of the bolded text from all of the exercises.

Bolded readings from these exercises are listed below as a summary review.

Pressure = Flow x Resistance

Power = Flow x Pressure

More flow = More speed

Lowering the pressure of a fluid = Less force

Raising the pressure of a fluid = More force

A weighted platen can lower a load using a single acting cylinder and gravity.

If you want to adjust the speed of a cylinder, then go for the flow control valve.

If you want to adjust the force of a cylinder, then go for the pressure relief valve.

A double acting cylinder can be mounted in any position.

Pressure is usually given in PSI, bar, or kPa.

If you want to adjust the speed of an actuator, then go for the flow control valve.

If you want to adjust the speed of a motor, then go for the flow control valve.

If you want to adjust the torque of a motor at constant flow, then go for the pressure relief valve.

As an actuator moves, the pressure at the input port and also the pressure against the cap end side of the piston wall of the cylinder is around 200 PSI [1378 kPa] or less.

As a cylinder reaches the clamp point (or as it is almost fully extended), the pressure at the input port and also the pressure against the cap end side of the cylinder's piston wall is heading to line pressure.

As the external pilot line of a pressure control valve is moved around, the pressure trip point changes.

The external pilot line can generally have one of three conditions. Send it to tank, send it another pressure relief valve or circuit element, or block it.

Any external pilot line pressure to a pressure relief valve lower in value than the mechanically set trip point of that pressure relief valve, and above the value of the unloaded pressure relief valve, will override the set point and cause that pressure relief valve to actuate.

If you have a fire, (1) get people out of harm's way, (2) report or call the fire in next, and (3) fight the fire if you are allowed to in your work situation.

III. Do the regular exercises in this order:

List of topics covered

(1) Identify the units of primary and secondary treatment

(2) Determine the relationship between pressure and flow

(3) Temperature and viscosity

(4) Double acting cylinders and 4-way directional control valves

(5) The relationship $F = P\,A$

- (6) Single acting cylinder and 3-way valve with suspended load
- (7) Meter in and meter out
- (10) Cam operated directional control valve to demonstrate additional aspects of flow control circuitry
- (11) Hydraulic motor
- (13) Sequence valve to demonstrate step control circuitry for a clamp and drill operation
- (14) Pressure reducing valve
- (15) Three trigger point pressure relief valve
- (16) Regenerative circuit

IV. Do the no build and optional exercises if desired:
- (8) No build—from print only—sequencing
- (9) OPTIONAL—No build—from print only—air and oil time delay
- (12) OPTIONAL—No build—force and torque

Exercise 1

Objectives:

1) Identify the units of primary and secondary treatment that exist within a hydraulic machine or a hydraulic system.

2) Determine what types of "mechanical" solutions are installed in a hydraulic system and which ones are not present.

You need to find the biggest hydraulic unit that you can, either inside the buildings that you are working in, or on a piece of outdoor equipment. Do this either by yourself or with two or three other people. Answer the following questions and then discuss them with your instructor.

1) List the units of primary and secondary oil treatment that the hydraulic system has.

2) How many pumps or pump stages does this system have?

3) Where is the main drain for this system located?

4) Are there any types of "mechanical" solutions to the water problem in the headspace of the tank that you can find?

5) How much pitch do the lines of the hydraulic system have?

6) What is the capacity of the tank in cubic ft.? In cubic meters?

7) Within this hydraulic system, how many lines go back to tank?

8) Where is the low point of the system? The test points?

9) How much evidence of leakage are you finding with this system?

10) What do you think of the overall design of this unit?

11) Discuss your answers and observations with your instructor.

Exercise 2

Objective:

To determine the relationship between pressure and flow on the biggest hydraulic system that you have available to you.

The following two relationships are helpful to the understanding of hydraulics.

Pressure = Flow x Resistance
and
Power = Flow x Pressure

Flow is usually given in GPM [LPM or lpm], which must be changed to cubic feet [cubic meters]. GPM times .134 gives cubic feet per minute (1 cu.ft. = 1728 cu.in. and 1 gal. = 231 cu.in.). [LPM times .001 gives cubic meters per minute].

Pressure is usually given in PSI [kPa] which must be changed to lbs./sq.ft. or PSF [Pa]. Multiplying PSI by 144 gives PSF [kPa times 1000 gives Pascals or Pa].

Power is usually given in horsepower or HP [kW], which must be changed to foot pounds per second [newton meters per second].

1 HP = 550 ft.lbs./sec. = 746 watts = [746 newton meters/sec.]

Pressure is equal to flow times resistance. This general relationship holds true for hydraulics (liquid pressure), pneumatics (gas pressure), and electricity (voltage pressure). In electricity this relationship is known as Ohm's law and is written:

E = I x R

Electrical power is usually expressed in watts, which is calculated using this formula:

Power = Flow x Pressure

(watts) = (amps) x (volts)

Watts can then be converted to horsepower [kW] as 746 watts = 1 HP or .746 kW = 1 HP.

These examples from the electrical world are included primarily as illustrations of the general relationship between pressure, flow, and resistance. The two formulas at the top of the page are the ones that you need to be familiar with. All of these subjects would be covered in greater depth in a fluid mechanics course.

You are going to find the biggest hydraulic unit that you can, either inside the building that you are working in, or on a piece of outdoors equipment. Do this either by yourself or with two or three other people. Answer the following questions:

1) How much flow can the pump provide?

2) What is the system pressure?

3) How much electrical power does the pump use?

4) What does GPM stand for? LPM?

5) What are four standard conversions for horsepower?

6) What is the capacity of the tank in cubic feet? in cubic meters?

7) What is the main pressure relief valve set at?

8) Discuss your answers and observations with your instructor.

Exercise 3

Objective

Determine the relationship between pressure, flow, temperature, and viscosity on a hydraulic system.

The following four relationships are helpful to the understanding of hydraulics, whichever way you say it.

Lowering the viscosity of a fluid = Slightly more flow

Heating a fluid = "Thinner" fluid

"Thinner" fluid = Slightly more flow

More flow = More speed

Flow is usually given in GPM [LPM], which is usually changed to cubic feet [cubic meters] for formulas. GPM times .134 gives cubic feet per minute [LPM times .001 gives cubic meters per minute] for formulas.

Pressure is usually given in PSI [kPa], which must be changed to lbs./sq.ft. or PSF [Pa] for formulas. Multiplying PSI by 144 gives PSF [kPa times 1000 gives Pa] for formulas.

You are going to build the following hydraulic circuit:

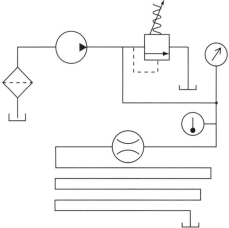

Use lots of right angle turns in this circuit and the smallest diameter hoses that you have. You want to be able to create as much heat in this circuit as possible. Run this circuit for at least an hour and a half, taking pressure, temperature, and flow readings at the beginning, the middle, and the end of the allotted time period. You can send everyone to break as long as one person stays with the running circuit.

You should be able to find all of these parts inside the building that you are working in if you are at a manufacturing plant site or in the stores area of your company if you work on outdoor equipment. The third option is to purchase any pieces from a local or regional hydraulic parts distributor. Do this exercise either by yourself or with another three or four other people. You can use either standard or metric units. Answer the following questions:

1) How much flow does the pump provide at the beginning and the end of this exercise?

2) What is the system pressure at the beginning and the end of this exercise?

3) How much temperature rise did you get from the beginning to the end of this exercise?

4) What does GPM stand for? LPM?

5) What are two or three additional ways to state some of the first four relationships listed at the beginning of this exercise?

6) What is the capacity of your tank in gallons? In liters?

7) What is the main pressure relief valve set at?

8) What is the percentage increase in the flow rate over the course of the exercise?

9) What is the percentage decrease in the pressure drop over the course of the exercise?

10) What is the percentage increase in the temperature of the hydraulic fluid over the course of the exercise?

11) Do you see why folks might consider using temperature or pressure compensated flow control valves on more critical or more closely controlled hydraulic circuits?

12) Discuss your answers and observations with your instructor.

Exercise 4

Objective:

Gain a hands-on understanding of double acting cylinders and 4-way directional control valves. Determine the relationship between pressure, flow, and speed.

The following four relationships are helpful to the understanding of hydraulics, whichever way you say it.

Lowering the pressure of a fluid = Less force

Less force (linear motion) = Less torque (rotary motion)

Lowering the pressure of a fluid = Slightly less flow

Slightly less flow = Slightly less speed

Flow is usually given in GPM [LPM], which is changed to cubic feet [cubic meters] for formulas. GPM times .134 gives cubic ft. per minute [LPM times .001 gets cubic meters per minute] for formulas.

Pressure is usually given in PSI [kPa], which is changed to lbs./sq.ft. or PSF [Pa] for formulas. Multiplying PSI by 144 gives PSF [kPa times 1000 gets Pa] for formulas.

Construct a circuit from the print below, then follow the instructions and answer the questions on the following page.

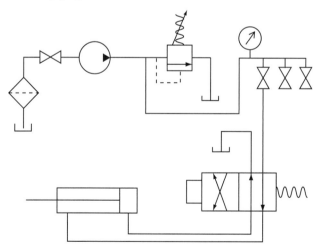

You should be able to find all of these parts inside the building that you are working in if you are at a manufacturing plant site or in the stores area of your company if you work on outdoor equipment. Purchase any other pieces needed from a hydraulic parts distributor. Do this exercise either by yourself or with one other person. You can use either standard or metric units.

A) Turn on the pump and set the system pressure at 500 PISG [3447 kPa].

B) Open the manifold supply valve.

C) Activate and hold the directional control valve in the "on" position.

  (1) What happens?

D) Release the trigger of the directional control valve.

  (2) What happens?

E) Toggle the trigger of the directional control valve several times.

  (3) What happens?

F) Reduce the system pressure by 30%.

G) Toggle the trigger of the directional control valve several times.

  (4) Do you notice any difference in the speed of extension and retraction?

  (5) What position is the cylinder in when the directional control valve is in the rest position?

  (6) What position is the cylinder in when the directional control valve is in the activated position?

  (7) What would the circuit look like if it was electrically operated and the cylinder was extended when the directional control valve was in the rest position?

  (8) How many ports does the directional control valve have?

  (9) Could you use a three ported directional control valve to extend and retract a double acting cylinder?

H) Disassemble the circuit.

I) Discuss your answers and observations with your instructor.

Exercise 5

Objective

Gain a better understanding of the relationship

F = P A

Where F = force in lbs. [newtons]
P = pressure in PSI [Pa]
A = area in sq. inches [sq. meters]

The following two relationships are helpful to the understanding of hydraulics.

Lowering the pressure of a fluid = Less force

Raising the pressure of a fluid = More force

Pressure is usually given in PSI [kPa], which usually is changed to lbs./sq.ft. or PSF [Pa] for formulas. Multiplying PSI by 144 gives PSF [kPa times 1000 gives Pa] for formulas.

Do this exercise either by yourself or with two or three other people. You can use either standard or metric units.

A) Find the location of a car lift in a local auto shop. Obtain permission to do the following experiment.

B) Raise the car lift to its full height with no load on it.

C) Measure the circumference or the radius of the main upright lift support (the rod).

D) Find out the operating pressure of the compressed air over oil system that runs the lift.

E) Lower the lift.

(1) Determine the maximum weight that the car lift can raise, using the relationship F = P A.

Almost all hydraulic car lifts are rams. This means that the diameter of the rod and the diameter of the piston are the same.

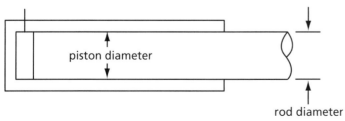

F) Discuss your answers and observations with your instructor.

Exercise 6

Objective

Use a hydraulic circuit that has a manually operated 3 way directional control valve with a suspended load to investigate the operation of this type of directional control valve and single acting hydraulic cylinders with suspended loads.

The following two relationships are helpful to the understanding of hydraulics.

A weighted platen can lower a load using a single acting cylinder and gravity.

A double acting cylinder can be mounted in any position.

Pressure is usually given in PSI or kPa. You can use either standard or metric units.

Construct this circuit and then answer the following questions:

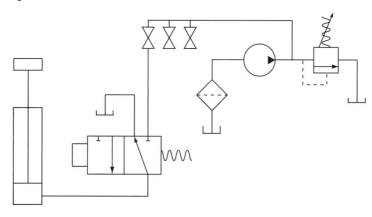

A) On the threaded rod end of the single acting cylinder, install two large nuts.

B) Attach a heavy plate of scrap metal to these two nuts and the rod to create a platen.

C) Mount the single acting cylinder in an upright position.

D) Put a small bag of concrete mix or other weight 80 lbs. or more [36 kg. or more—it really is 360 newtons or more] on top of the platen.

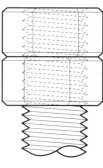

side view

E) Turn on the pump and set the pressure to 200 PSIG [1378 kPa].

F) Open the manifold valve.

G) Slowly toggle the trigger of the directional control valve several times.

H) Increase the pressure by 100 PSI [689 kPa].

I) Repeat steps G and H several times until the pressure reaches 900 PSIG [6205 kPa].

  (1) During this experiment, which quantity in the relationship of F = P A remained constant?

  (2) How much force was pressing against the platen when the pressure was 200 PSIG [1378 kPa]? 500 PSIG or 34 bar [3447 kPa]?

  (3) Did your platen ever get stuck? Why or why not?

J) Disassemble the circuit.

K) Discuss your answers and observations with your instructor.

Exercise 7

Objective

Use a hydraulic circuit that has a manually operated directional control valve to demonstrate aspects of flow control circuitry, such as meter in and meter out.

The following two relationships are helpful to the understanding of hydraulic actuators.

> **If you want to adjust the speed of a cylinder, then go for the flow control valve.**

> **If you want to adjust the force of a cylinder, then go for the pressure relief valve.**

Pressure is usually given in PSI or kPa. You can use either standard or metric units in this exercise.

Construct circuit #1 and answer the following questions:

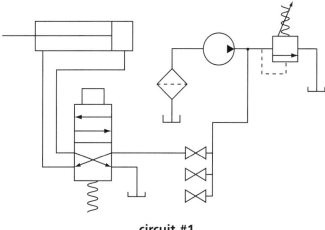

circuit #1

A) Install a variable flow control valve with integral check in this circuit to adjust the speed of retraction of the hydraulic cylinder.

B) Turn on the pump and set the hydraulic system pressure to 900 PSIG [6205 kPa].

C) Open the manifold valve.

D) Toggle the trigger of the directional control valve several times.

E) Adjust the needle valve in the flow control valve to various settings and repeat step (D).

   (1) Do you have full speed on retraction?

   (2) Is your circuit meter in or meter out?

   (3) Do you have full speed on extension?

F) Shut the manifold valve and slowly toggle the direction control valve a few times.

G) Install a variable flow control valve with integral check in this circuit to adjust the speed of extension of the cylinder.

H) Repeat steps C through E and question 2.

I) Sketch in the print of circuit #1 the positions and the orientations of your flow control valves.

   (4) Does the circuit look like circuit #2 at the bottom of the page?

   (5) Does your circuit accomplish the same thing as circuit #2 would?

J) Disassemble or rework your circuit and construct circuit #2.

K) Repeat steps B through E and questions 1 through 3.

L) Please note that the flow control valve orientation in circuit #2 (meter out) is the most common in

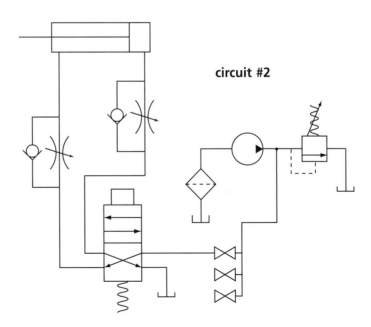

circuit #2

hydraulics. Whatever school that you and your co-workers are involved in (meter in or meter out) is okay. You need to recognize and troubleshoot the situation that you happen to encounter on the machine or system that you will be working on.

M) Disassemble the circuit.
N) Discuss your answers and observations with your instructor.

Exercise 8
(no build—from print only)

Objective:

Use a hydraulic circuit print that has both solenoid operated and hydraulic pilot operated directional control valves to demonstrate sequencing or time delay circuitry.

Using the circuit below (from Self-Test #2), answer the following questions:

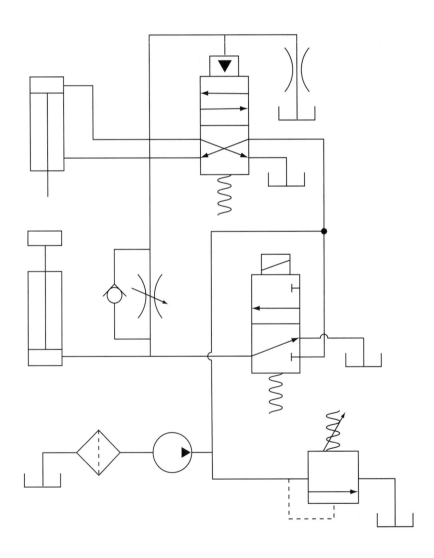

The following two relationships are helpful to the understanding of hydraulics.

A weighted platen can lower a load using a single acting cylinder.

A double acting cylinder can be mounted in any position.

Pressure is usually given in PSI or kPa. You can use either standard or metric units.

(no build—from print only)

A) Hook up a solenoid to the appropriate voltage source as follows:

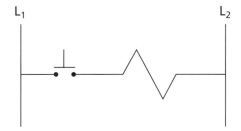

B) Turn on the pump and set the pressure to 900 PSIG [6205 kPa].

C) Open the manifold valve.

D) Slowly toggle the pushbutton switch several times.

E) Adjust the needle valve to obtain a time delay of five seconds.

F) Slowly toggle the pushbutton switch several times.

G) Adjust the needle valve to obtain a time delay of fifteen seconds between the two cylinder extensions.

   (1) This circuit can be used to create a time delay in a hydraulic circuit. Usually solid state electronic circuits or PLCs are used for time delays. Do you know why?

   (2) Why is the electrical hookup not on the hydraulics print?

   (3) Why is the dotted line attached to a flow control valve?

(4) If in addition to the time delay you wanted to independently control the speed of extension and retraction of all of the cylinders in this circuit, then how many flow control valves with integral checks would you use?

(5) Sketch on the print in the proper position and orientation all of the flow control valves mentioned in question #4.

H) Disassemble the circuit.

I) Discuss your answers and observations with your instructor.

Exercise 9 (optional)
(no build—from print only)

Objective

Use an integrated fluid power circuit (pneumatics and hydraulics) to investigate the types of components and fittings used to build hydraulic circuits.

The following two relationships are helpful to the understanding of fluid power.

> **Pressure is usually given in PSI, bar, or kPa.**
>
> **A single acting cylinder with spring return in pneumatics can generally be mounted in any position.**

You can use either standard or metric units. Construct the circuit from the print on the next page. Make sure all the valves are closed and then answer the following questions:

(1) What would the list of the components that you would order for this circuit look like?

(2) Would your pressure gauge(s) be standard or metric?

A) Turn on the air compressor.

B) Open the hydraulic and the pneumatic shutoff valves.

C) Turn on the pump and set the hydraulic system pressure to 700 PSIG [4826 kPa].

(3) Would the double acting cylinder extend?

D) Set the pneumatic pressure to 50 PSIG [345 kPa].

E) Open the manifold valve.

(4) Would the single acting cylinder extend?

F) Toggle the trigger of the 3 way valve slowly several times.

(5) What would happen?

(6) How much force would be pressing against the piston face of the air cylinder when the pressure was 50 PSIG [345 kPa]?

(7) Would your double acting cylinder ever move? Why or why not?

G) Adjust the flow control valve if the answer to question 7 is, "The double acting cylinder never moved."

(8) Would that do the trick?

H) Disassemble the circuit.

I) Discuss your answers and observations with your instructor.

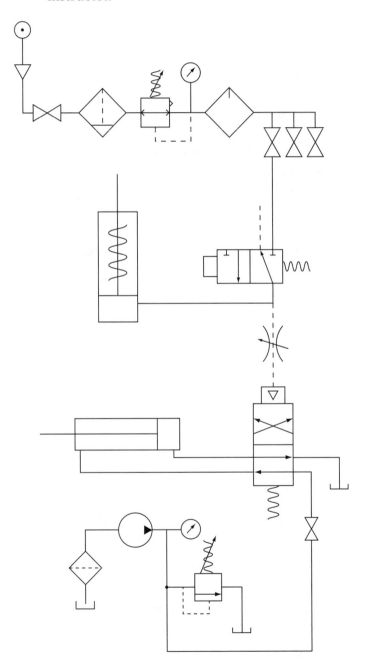

Exercise 10

Objective

Use a hydraulic circuit that has an electrically or manually operated directional control valve, and a cam operated directional control valve to demonstrate additional aspects of flow control circuitry, such as meter in or meter out, fast and/or slow.

The following three relationships are helpful to the understanding of hydraulic actuators.

If you want to adjust the speed of an actuator, then go for the flow control valve.

If you want to adjust the force of a cylinder, then go for the pressure relief valve.

If you want to adjust the timing of a cylinder, then go for the cam operated valve.

Pressure is usually given in PSI or kPa. You can use either standard or metric units in this exercise. Construct this circuit and answer the following questions:

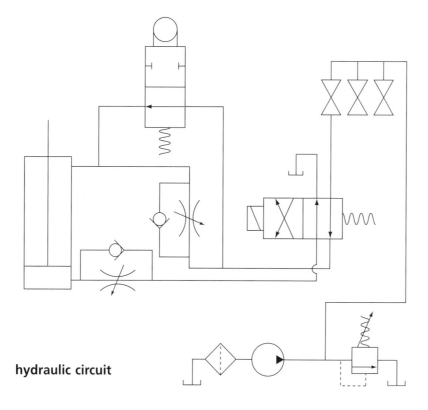

hydraulic circuit

A) Mount the cylinder and the 2 way valve so that the roller or cam is depressed halfway through the cylinder stroke.

B) Hook up the solenoid as follows to the appropriate voltage source.

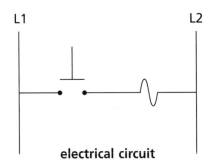

electrical circuit

C) Turn on the pump and set the pressure to 900 PSIG [6205 kPa].

D) Open the manifold valve.

E) Turn on the electricity for the solenoid control circuit.

F) Toggle the trigger of the pushbutton switch several times.

G) Adjust the needle valve(s) of the flow control valve(s) so that the cylinder extends slowly after it depresses the roller on the 2 way valve and the cylinder retracts at almost full speed.

H) Toggle the trigger of the pushbutton switch slowly several times to cycle the system.

(1) This circuit is called a rapid traverse and feed circuit, and is utilized primarily in machine shops. Do you know why?

(2) Why is the roller cam not placed next to the rod of the cylinder on the print?

(3) Why does the electrical print use L1 and L2?

(4) List the components that are in parallel with each other in this circuit.

(5) Are you metering in or metering out?

(6) How long does it take your circuit to make a complete cycle?

(7) How would this circuit work if you used a solenoid operated 2 way valve?

I) Shut the manifold valve and slowly toggle the direction control valve a few times.

J) Turn off the pump and the electricity.

K) Disassemble the circuit.

L) Discuss your answers and observations with your instructor.

Exercise 11

Objective

Use a hydraulic circuit that has a hydraulic motor to demonstrate flow control circuitry, such as meter in and meter out and control of freewheeling.

The following three relationships are helpful to the understanding of hydraulic motors.

> **If you want to adjust the speed of a motor, then go for the flow control valve.**

> **If you want to adjust the torque of a motor at constant flow, then go for the pressure relief valve.**

> **If you want to adjust the timing of a motor, then go for the electrically operated directional control valves.**

Pressure is usually given in PSI or kPa. You can use either standard or metric units in this exercise. Construct this circuit and answer the following questions:

hydraulic motor circuit

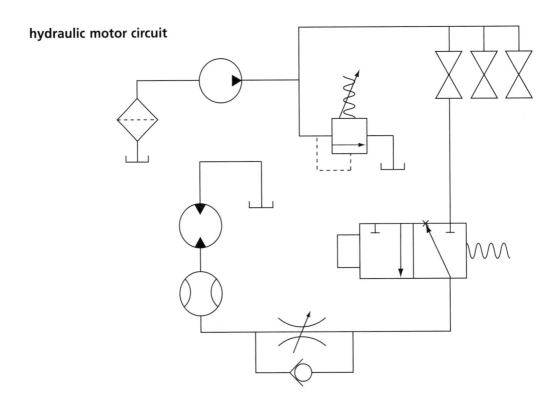

A) Turn on the pump and set the pressure to 200 PSIG [1378 kPa].
B) Open the manifold valve.
C) Adjust the needle valve on the flow control valve all the way out.
D) Toggle the trigger of the directional control valve.
E) Record the flow and RPM of the motor using the flow meter and a tachometer.
F) Increase the line pressure another 100 PSI [689 kPa].
G) Repeat steps D through F until 900 PSIG [6205 kPa] is reached.
H) Adjust the needle valve on the flow control valve one turn in.
I) Set the pressure to 200 PSIG [1378 kPa].
J) Toggle the trigger of the directional control valve.
K) Record the flow and RPM of the motor using the flow meter and a tachometer.
L) Increase the line pressure another 100 PSI [689 kPa].
M) Repeat steps J through L until 900 PSIG [6205 kPa] is reached.
N) Repeat steps H through M until the flow control valve is completely closed.
   (1) What is the relationship between pressure and RPM?
   (2) What is the relationship between flow and RPM?
   (3) What is another word for RPM?
   (4) What units is your flow meter rated in?
   (5) How much does your hydraulic motor weigh?
   (6) How does this weight compare to an electric motor of equal horsepower?
   (7) How long does it take your circuit to make a complete cycle?
   (8) How would this circuit work if you used a solenoid operated 2 way valve?

O) Turn off the pump.
P) Disassemble the circuit.
Q) Discuss your answers and observations with your instructor.

Exercise 12 (optional)

Objective

Use a hydraulic circuit that has a cylinder and a hydraulic motor to demonstrate independent pressure control circuitry for force and torque.

The following three relationships are helpful to the understanding of the hydraulic actuators in this exercise.

> **If you want to adjust the speed of an actuator, then go for the flow control valve.**
>
> **If you want to adjust the torque of a motor at constant flow, then go for the pressure relief valve.**
>
> **If you want to adjust the force of a cylinder, then go for the other pressure relief valve.**

Pressure is usually given in PSI or kPa. You can use either standard or metric units in this exercise. Construct the circuits and answer the following questions:

A) Construct the hydraulic circuit on the next page.

B) Construct either electrical circuit on page 161.

C) Turn on the pump and set the system pressure to 900 PSIG [6205 kPa]. Make sure the pressure relief valve for the motor is set halfway out from the setting of the main system pressure relief valve.

D) Turn on the electricity for the control circuit.

E) Open the manifold valve.

F) Adjust the needle valves on all of the flow control valves all the way out.

G) Toggle the trigger of only solenoid C side (the two way valve).

H) Record the flow and RPM of the motor using the flow meter and a tachometer.

I) Toggle the trigger of only the solenoid A side.

J) Toggle the trigger of only the solenoid B side.

K) Adjust the needle valves on all of the flow control valves halfway in.

L) Toggle the trigger of the cylinder's directional control valve (solenoid A side, then solenoid B side).

16 · Exercises

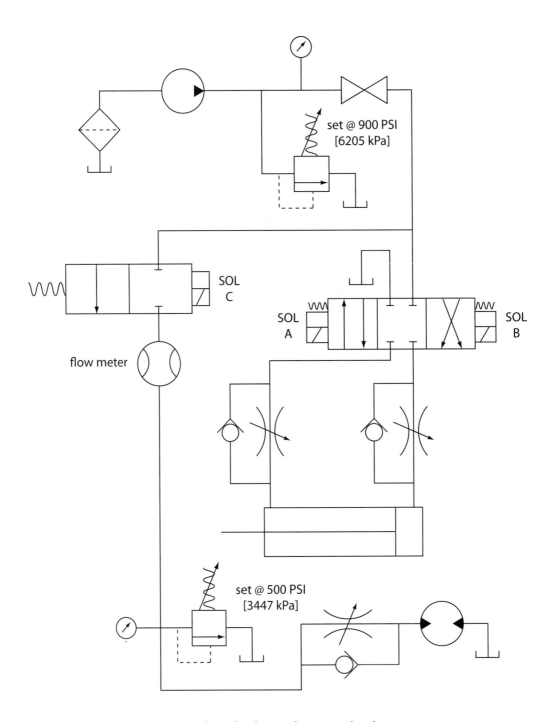

hydraulic cylinder and motor circuit

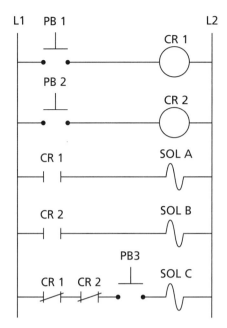

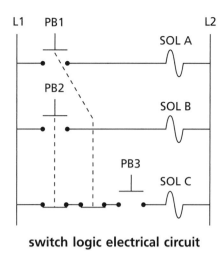

switch logic electrical circuit

control relay logic electrical circuit

M) Record the flow and RPM of the motor using the flow meter and a tachometer.

N) Repeat steps I through M while holding solenoid C side activated the whole time.

   (1) What is the relationship between A and B?
   (2) What is the relationship between A and B inverse, and C?
   (3) What is another word for the dotted line on the electrical print?
   (4) What units is your motor capacity rated in?
   (5) How much weight can your cylinder move?
   (6) How does this weight compare to the torque of your hydraulic motor?
   (7) How long does it take your cylinder to make a complete cycle if PB1 is depressed? If PB1 is not depressed?
   (8) With this particular logic (the solenoid C interlock), two pressure relief valves were used. If you did not use the solenoid C interlock,

then what two pressure control valves would you have to use for the circuit to work in the same manner?

(9) How would this circuit work if you used a PLC?

O) Turn off the pump and the electrical control circuit.

P) Disassemble the circuits.

Q) Discuss your answers and observations with your instructor.

Exercise 13

Objective

Use a hydraulic circuit that has two cylinders and a sequence valve in it to demonstrate step control circuitry for a clamp and drill operation.

The following relationships are helpful to the understanding of hydraulic actuators.

As an actuator moves, the pressure at the input port and also the pressure against the cap end side of the piston wall of the cylinder is around 200 PSI [1378 kPa] or so.

As a cylinder reaches the clamp point (or it is almost fully extended), the pressure at the input port and also the pressure against the cap end side of the piston wall of the cylinder is heading to system pressure. In this case that is 900 PSI [6205 kPa].

Pressure is usually given in PSI or kPa. You can use either standard or metric units in this exercise. Construct the two hydraulic circuits shown below and answer the following questions. Combine these two circuits through the use of a shutoff valve.

A) Construct the hydraulic circuit using either a manually operated or a solenoid operated directional control valve. Please remember that the rest position of the directional control valve is the center position at PTAB.

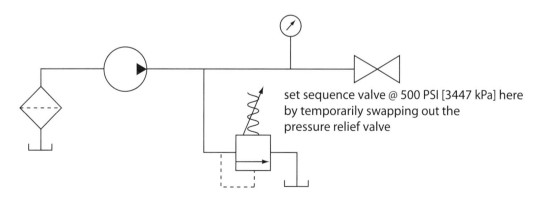

set sequence valve @ 500 PSI [3447 kPa] here by temporarily swapping out the pressure relief valve

pump circuit

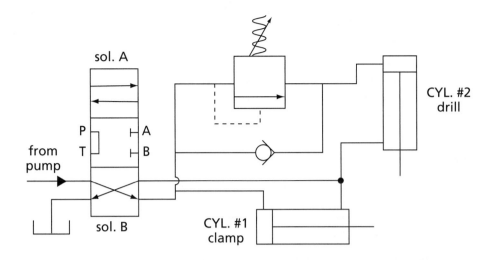

B) Construct the electrical circuit needed if you are using a solenoid operated directional control valve.

C) Turn on the pump and set the system pressure to 900 PSIG [6205 kPa]. Make sure the sequence valve for the #2 cylinder (drill) has been preset at 500 PSI [3447 kPa].

D) Turn on the electricity for the control circuit if used.

E) Open the shutoff valve.

F) Adjust the needle valves on any of the flow control valves (that you add to this circuit) all the way out.

G) Toggle the trigger of the directional control valve to the right (use only solenoid A if using an electric operated valve).

H) Record the timing and speed of the cylinders using a watch and observation.

I) Toggle the trigger of the directional control valve to the left (use only solenoid B if using an electric operated valve).

J) Adjust the needle valves on any flow control valves halfway in.

K) Toggle the trigger of the cylinder's directional control valve to the right and then to the left (solenoid A side, then solenoid B side).

L) Record the timing and speed of the cylinders using a watch and observation.

   (1) What is the timing relationship between cylinder #1 (clamp) and cylinder #2 (drill)?

   (2) What is the speed relationship between cylinder #1 and #2?

   (3) What is another word for this circuit?

   (4) What units is your pump's volumetric capacity rated in?

   (5) How much force can your cylinders push with?

   (6) How would this pushing force compare to the same circuit with a system pressure of 700 PSI [4826 kPa]?

   (7) How would this circuit work if you wanted to push on stuff faster and drill quicker?

M) Turn off the pump (and the electrical control circuit if used).

N) Disassemble the circuit(s).

O) Discuss your answers and observations with your instructor.

16 · Exercises

Exercise 14

Objective

Use a hydraulic circuit that has two cylinders and a pressure reducing valve in it to demonstrate the control circuitry and the working circuit for a clamping operation.

The following relationships are helpful to the understanding of hydraulic actuators.

As an actuator moves, the pressure at the input port and also the pressure against the cap end side of the piston wall of the cylinder is around 200 PSI [1378 kPa] or less.

As a cylinder reaches the clamp point (or as it is almost fully extended), the pressure at the input port and also the pressure against the cap end side of the cylinder's piston wall is heading to line pressure. In this case that is 500 PSI [3447 kPa] for cylinder B and 900 PSI [6205 kPa] for cylinder A.

Pressure is usually given in PSI or kPa. You can use either standard or metric units in this exercise. Construct the two circuits shown below, then combine these two circuits through the use of a shutoff valve and add a pressure gauge at or near the cap end of both cylinders.

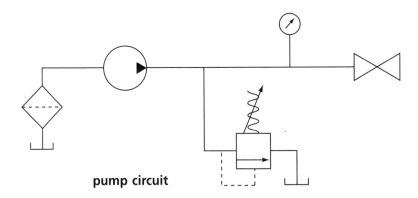

pump circuit

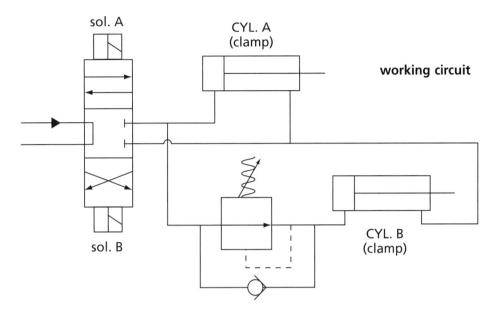

A) Construct the hydraulic circuit using either a manually operated or a solenoid operated directional control valve.

B) Construct the electrical circuit needed if you are using a solenoid operated directional control valve.

C) Turn on the pump and set the system pressure to 900 PSIG [6205 kPa]. Make sure the pressure reducing valve for the B cylinder has been preset at 500 PSI [3447 kPa].

D) Turn on the electricity for the control circuit if used.

E) Open the shutoff valve.

F) Adjust the needle valves on any of the flow control valves (that you add to this circuit) all the way out.

G) Toggle the trigger of the directional control valve to the right (use only solenoid A if using an electric operated valve).

H) Record the pressure and speed of the cylinders using a watch and pressure gauges.

I) Toggle the trigger of the directional control valve to the left (use only solenoid B if using an electric operated valve).

J) Adjust the needle valves on any flow control valves halfway in.

K) Toggle the trigger of the cylinder's directional control valve to the right and then to the left (solenoid A side, then solenoid B side).

L) Record the time and pressure of the cylinders using a watch and pressure gauges.

   (1) What is the time relationship between cylinders A and B?

   (2) What is the pressure relationship between cylindera A and B?

   (3) What is another word for this circuit?

   (4) What units is your pump capacity rated in?

   (5) How much force can your cylinders clamp with?

   (6) How would this clamping force compare to the same circuit with a system pressure of 700 PSI [4826 kPa]?

   (7) How would this circuit work if you wanted to clamp faster?

M) Turn off the pump (and the electrical control circuit if used).

N) Disassemble the circuit(s).

O) Discuss your answers and observations with your instructor.

Exercise 15

Objective

Use a hydraulic circuit that has a system shutoff valve and pressure relief valves in it to demonstrate the control circuitry and the working circuit for a three pressure operation system.

The following relationships are helpful to the understanding of hydraulic system pressures.

As the external pilot line of a pressure control valve is moved around, the pressure trip point of that pressure control valve changes.

The external pilot line can generally have one of three conditions. Send it to tank, send it another pressure relief valve or circuit element, or block it. In this case that is about 150 PSI [1034 kPa], 700 PSI [4826 kPa] and 1100 PSI [7584 kPa], respectively.

Pressure is usually given in PSI or kPa. You can use either standard or metric units in this exercise. Use the following illustration for your working circuit.

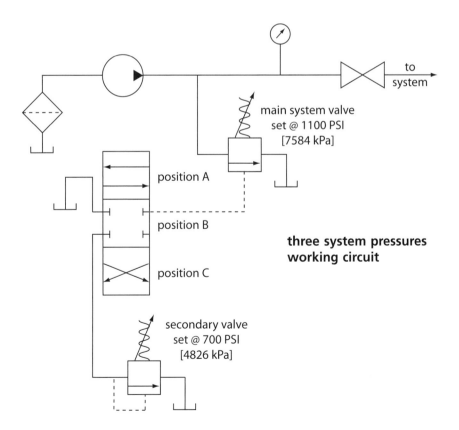

three system pressures working circuit

Make sure that you add a pressure gauge in front of the system shutoff valve.

A) Construct the hydraulic circuit using either a manually operated or a solenoid operated directional control valve.

B) Construct the electrical circuit needed if you are using a solenoid operated directional control valve.

C) Turn on the pump and set the system pressure to 1100 PSIG [7584 kPa]. Make sure that the second pressure relief valve for the secondary system setting has been preset at 700 PSI [4826 kPa].

D) Turn on the electricity for the control circuit if used.

E) Keep the shutoff valve CLOSED.

F) Record the pressure of the system when the remote pilot of the main pressure relief valve is blocked (position B—the rest position).

G) Toggle the trigger of the directional control valve to the left for position C (use only the solenoid for position C if using an electric operated valve).

H) Record the pressure when the remote pilot of the main pressure relief valve is sent to the secondary pressure relief valve (position C—the override position).

I) Toggle the trigger of the directional control valve to the right for position A (use only the solenoid for position A if using an electric operated valve).

J) Record the pressure when the remote pilot of the main pressure relief valve is sent to tank (position A—the unloaded position).

K) Record the system pressure of each of the three positions and compare.

    (1) What is the pressure relationship between position A and C?

    (2) What is the pressure relationship between position A and B?

    (3) What is another name for this circuit?

- (4) What units is your pump's electrical capacity rated in?
- (5) How much pressure can your cylinders clamp with in position B?
- (6) How would this clamping force compare to the same circuit using position C?
- (7) How would this circuit work if you wanted to unload the pump?

L) Turn off the pump (and the electrical control circuit if used).

M) Disassemble the circuit.

N) Discuss your answers and observations with your instructor.

Exercise 16

Objective

Use a hydraulic circuit that has a cylinder, a pressure relief valve, and a float center directional control valve in it to demonstrate the control circuitry and the working circuit for a regenerative circuit (a rapid extension operation).

The following position relationships are helpful to the understanding of hydraulic regeneration circuits.

Position #1—retract at regular speed

Position #2—extend at fast speed at a much reduced clamping force

Position #3—extend at regular speed at full clamping force

Pressure is usually given in PSI or kPa. You can use either standard or metric units in this exercise. Use the illustrations below for your pump circuit and your working circuit. Combine these two circuits through the use of a shutoff valve and use the system pressure gauge in the port P line.

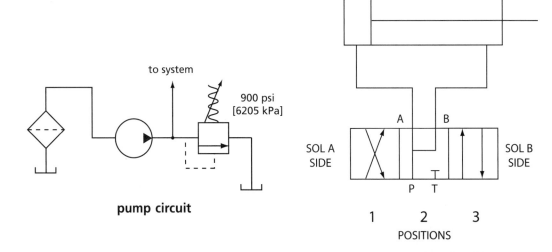

pump circuit

regenerative circuit
(working circuit)

A) Construct this hydraulic circuit using either a manually operated or a solenoid operated directional control valve.

B) Construct the electrical circuit needed if you are using a solenoid operated directional control valve.

C) Turn on the pump and set the system pressure to 900 PSIG [6205 kPa].

D) Turn on the electricity for the control circuit if used.

E) Open the shutoff valve.

F) Adjust the needle valves on any of the flow control valves (that you add to this circuit) all the way out.

Step 1—regular speed on extension

G) Toggle the trigger of the directional control valve to the left to position #1 (use only solenoid A if using an electric operated valve).

H) Record the pressure and speed of the cylinder on retraction using a watch and the system pressure gauge.

I) Toggle the trigger of the directional control valve to the right to position #3 (use only solenoid B if using an electric operated valve).

J) Toggle the trigger of the cylinder's directional control valve to the left to position #1 and then to the right to position #3 (solenoid A side then solenoid B side).

K) Record the time and pressure of the cylinder on retraction and extension using a watch and pressure gauges.

Step 2—fast speed on extension

L) Toggle the trigger of the directional control valve to the left to position #1 (use only solenoid A if using an electric operated valve).

M) Record the pressure and speed of the cylinder on retraction using a watch and the system pressure gauge.

N) Toggle the trigger of the directional control valve to the middle position #2 (do not use solenoid A and B if using an electric operated valve).

O) Toggle the trigger of the cylinder's directional control valve to the left to position #1 and then to the middle position #2 (solenoid A side then to the no solenoid middle position).

P) Record the time and pressure of the cylinder on retraction and extension using a watch and the system pressure gauge.

   (1) What is the time relationship between solenoid sides A and B (positions 1 and 3)?

   (2) What is the pressure relationship between solenoid sides A and B (positions 1 and 3)?

   (3) What is the time relationship between solenoid side A (position #1) and the no solenoid activated middle (position #2)?

   (4) What is the pressure relationship between solenoid side A (position #1) and the no solenoid activated middle (position #2)?

   (5) What is another name for this circuit?

   (6) What units is your cylinder capacity rated in?

   (7) What is the maximum force that your cylinder can clamp with?

   (8) Describe what happens at position #1.

   (9) Describe what happens at position #2.

   (10) Describe what happens at position #3.

Q) Are your explanations similar to those listed below for positions 1-3?

R) Turn off the pump (and the electrical control circuit if used).

S) Disassemble the circuit(s).

T) Discuss your answers and observations with your instructor.

Position #1: The pressure at port B and also the pressure against the rod end side of the cylinder's piston wall will be around 200 PSI [1378 kPa] or less as the cylinder is retracting. The pressure at port B and also the pressure against the rod end side of the cylinder's piston wall will then head to line pressure as the cylinder fully retracts. In this case that is 900 PSI [6205 kPa] for the cylinder.

Position #2: As the cylinder extends in a regenerative circuit, the fluid from the pump and from the rod side of the cylinder both enter the cap side of the cylinder. The output force of the cylinder is very low because the pressures against the cap end side and the rod end side of the cylinder piston are almost cancelling each other out. The only difference in force is the surface area equivalent of the piston rod times pressure pushing on the cap end side of the cylinder piston.

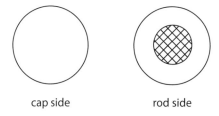

cap side   rod side

cylinder piston face areas

Position #3: As a cylinder nears the clamp point, the cylinder must be taken out of regeneration mode and put back into regular extension mode. The pressure at port A and also the pressure against the cap end side of the cylinder's piston wall will then head to line pressure. In this case that is 900 PSI [6205 kPa] for the cylinder.